iPhone 12 User Guide

A detailed Manual to understand the new iPhone 12 for Beginners, and professionals with hidden tricks, and Short Cut Keys

by

Brian B. Mooney

Copyright © 2020 Brian B. Mooney

All rights reserved. No part of this book shall be reproduced, stored in a retrieval system, or transmitted by any means, electronic, mechanical, photocopying, recording, or otherwise, without written permission from the publisher. Although every precaution has been taken in the preparation of this book, the publisher and author assume no responsibility for errors or omissions. Nor is any liability assumed for damages resulting from the use of the information contained herein.

Table of Contents

iPhone 12 1
User Guide 1
INTRODUCTION 12
CHAPTER ONE 1
TURN ON AND CONFIGURE IPHONE 1
   Prepare for setup.. 1
   Turn on and configure iPhone... 2
   Switch from Android device to iPhone............................. 4
   Set up mobile service on iPhone 6
   Install the nano-SIM card... 7
   Remove the account from iPhone..................................... 8
   Set up your mobile package using eSIM........................... 9
CHAPTER TWO.. 11
MANAGE YOUR MOBILE PLANS....................... 11
   When using a dual SIM, Do The following:..................... 12
CHAPTER THREE 13
CONNECT THE IPHONE TO THE INTERNET 13
   Join a personal hotspot 13
   Connect iPhone to a mobile network............................... 14
CHAPTER FOUR 16
MANAGE APPLE ID AND ICLOUD SETTINGS ON IPHONE 16

Sign in with your Apple ID 17

Change iCloud settings 18

Ways to use iCloud on iPhone.. 19

Wake up and unlock iPhone 20

CHAPTER FIVE..21
WAKE UP IPHONE 21

Unlock iPhone with Face ID 23

Unlock iPhone with Touch ID... 24

Unlock iPhone with a password 24

CHAPTER SIX.. 26
LEARN THE BASIC GESTURES OF INTERACTING WITH YOUR IPHONE 26

Adjust the iPhone volume 27

Lock the ring volume and alarm volume in Settings 28

Adjust the volume in the control center 28

Limits the headphone volume 28

Set iPhone to call or silent mode 30

CHAPTER SEVEN 33
SWAP THE SOUNDS AND VIBRATIONS OF THE IPHONE 33

Adjust the sound and vibration options............................ 33

Turn haptic feedback on or off.. 34

CHAPTER EIGHT.. 35

FIND IPHONE SETTINGS **35**

CHAPTER NINE **38**

ARRANGE THE BRIGHTNESS AND COLOR OF THE IPHONE SCREEN **38**

   Turn dark mode on or off 38

   Set dark mode to turn on and off automatically 40

   Adjust the screen brightness manually 40

   Turn True Tone on or off 41

   Tap Off, then select Sunset to Sunrise or Custom Plan. 42

CHAPTER TEN **43**

ENLARGE IPHONE SCREEN WITH SCREEN ZOOM **43**

Change the iPhone name **43**

Adjust the date and time on the iPhone **44**

Enter language and region on iPhone **45**

Set up email, contacts, and calendar accounts .. **45**

   Create a calendar account 46

Access features on the iPhone lock screen **47**

   Access features and information with Screen Lock 48

   To select information available from the lock screen, see Checking Access to Information on the iPhone Lock Screen. 49

Open apps on iPhone **50**

   Open applications on the home screen 50

Explore the app library 53

Hide and show pages on the home screen 56

Take a screenshot or screenshot on iPhone 57

Take a screenshot 57

Create a screenshot 58

Change or lock the screen orientation on iPhone 59

Lock or unlock screen orientation 60

Change iPhone wallpaper 61

Set a Live Photo as the background for the lock screen 63

Switch between iPhone apps 64

Use Application Switcher 64

Switch between open applications 66

Multitasking picture in picture on iPhone 66

Type using the on-screen keyboard on iPhone .. 68

Enter text using the on-screen keyboard 68

Make the on-screen keyboard a trackpad 72

Enter letters with accents or other characters as you type 73

Move text 75

Specify write settings 77

Enter with one hand 78

CHAPTER ELEVEN 79
USE AIRDROP 79

Before you begin 79

Using AirDrop 80

How to accept AirDrop 82

Change AirDrop settings 84

CHAPTER TWELVE 87
CHANGE NOTIFICATION ON IPHONE 12 87

Grouped alerts 87

Instant setting 89

Critical alerts 90

Number of alerts during screen time 93

Richer alerts 95

Switch on Do not disturb while driving 97

You can receive calls, messages, and alerts when you are a passenger 98

Send an auto-reply message as you drive 99

Create a custom auto-reply message 100

Allow some calls 100

Dictated text on iPhone 100

Turn on dictation 101

Dictation of text 101

Format punctuation or text 103

CHAPTER THIRTEEN 105
ADD OR CHANGE KEYBOARDS ON IPHONE .. 105

Add or remove a keyboard for another language 105

Switch to another keyboard 106

Assign an alternate layout to the keyboard 107

Use the iPhone to search 107

You can choose which apps you want to include in the search results 107

Search on iPhone 108

Turn off search suggestions 109

Search applications 110

Add a dictionary 110

CHAPTER FOURTEEN 111

CUSTOMIZE THE CONTROL CENTER ON IPHONE 111

Customize the control center 112

Add widgets to the iPhone Home screen 115

Open today's view 116

Move a device from the day view to the home screen.... 116

Add a gadget to the Home screen 117

Edit module 117

Remove a device from the Home screen 118

Give access to the day view when iPhone is locked 118

Select iPhone settings to travel 119

Turn on airplane mode 119

In airplane mode, turn on Wi-Fi or Bluetooth 120

Use screen time on iPhone, iPad, or iPod touch
.... 122

Turn on screen time.. 123

Enter a screen time password... 125

See your report and set boundaries................................. 126

Downtime 128

App limits 129

Communication boundaries... 129

always allowed 129

Content and privacy restrictions 130

Spend screen time with your family................................ 130

CHAPTER FIFTEEN 132

SET CONTENT AND PRIVACY RESTRICTIONS 132

Prevent purchases from iTunes and the App Store 133

Enable built-in applications and features........................ 134

Prevent explicit content and content rating.................... 136

Prevent web content 137

Restrict Siri Internet Search 139

Allow changing the privacy settings................................. 141

Allow changes to other settings and services.................. 142

CHAPTER SIXTEEN 144

ASK SIRI ON IPHONE.. 144

Invite Siri with your voice 144

Call Siri with a button 145

Correct if Siri is misunderstood 146

Write instead of talking to Siri 146

Find out what Siri can do on the iPhone 147

CHAPTER SEVENTEEN 150
SHARE YOUR VOICE WITH IPHONE AIRPODS AND BEATS 150

Share your voice with your friend Beats headphones 151

Use AirPods or a Beats product. 153

Adjust the volume of each toolkit 153

Stop sharing audio 154

CHAPTER EIGHTEEN 156
COLLECT HEALTH AND FITNESS DATA IN THE IPHONE APP FOR IPHONE 156

Manually update your health profile 157

Touch Health Details, then touch Edit. 159

Collect data from other sources 160

Set reminders on iPhone 161

Keep reminders up to date on all your devices using iCloud 162

Add a reminder 162

Move or delete reminders 164

CHAPTER NINTEEN TAKE PICTURES WITH YOUR IPHONE CAMERA 166

Take a picture 167

Turn the flash on or off 167

Set a timer 168

Zoom in or out 168

Laugh and selfie 168

Adjust the camera's focus and exposure 169

Take pictures in low light with night mode 170

Take a live photo 171

Take a panoramic picture 172

Take a picture with a filter 172

Take continuous pictures 173

INTRODUCTION

Once a year, Apple launches a major update to all iOS software running on iPhones. Millions of devices get free new features, along with security bug updates and interface changes. The latest version of iOS was announced at WWDC on June 22nd and became available on Wednesday, September 16th.

IOS 14 has been updated a little over a week after its launch, and now from September 24, version 14.0.1, with more details on the fixes shown below.

In this article, we'll cover everything you need to know about iOS 14, including any issues and issues with the latest version that iPhones may run on the new operating system, as well as all the new features and functionality that are still under development. We also have details on the developer and public beta versions and provide insight into new features that are still available on iPhones running iOS 14.

For a comparison with the previous iPhone operating system, see iOS 14 vs iOS 13. We have an overview of iOS 14 that is worth reading.

CHAPTER ONE

TURN ON AND CONFIGURE IPHONE

Turn on and configure your new iPhone over an Internet connection. You can also set up the iPhone by connecting it to your computer. If you have another iPhone, iPad, iPod touch, or Android, you can transfer data to your new iPhone.

If your iPhone is distributed or managed by a company or other organization, you can find installation instructions from your administrator. For general information, visit the Apple at Work website.

Prepare for setup

The following things are available for the smoothest possible layout:

- Internet connection via Wi-Fi network (network name and password may be required) or mobile data service through one operator

- Apple ID and password; If you do not have an Apple ID, you can create one during installation
- Credit or debit card information if you want to add a card to Apple Pay during setup
- A backup of your previous iPhone or device when you transfer data to your new device
- On your Android device by transferring your Android content

Turn on and configure iPhone

Press and hold the side button or the Sleep / Wake button (depending on the model) until the Apple logo display on the screen

Side or Sleep / Wake button on three different models of iPhone

If the iPhone doesn't turn on, you may need to charge the battery. See the Apple Support article for more help. If your iPhone, iPad, or iPod touch doesn't turn on or freezes.

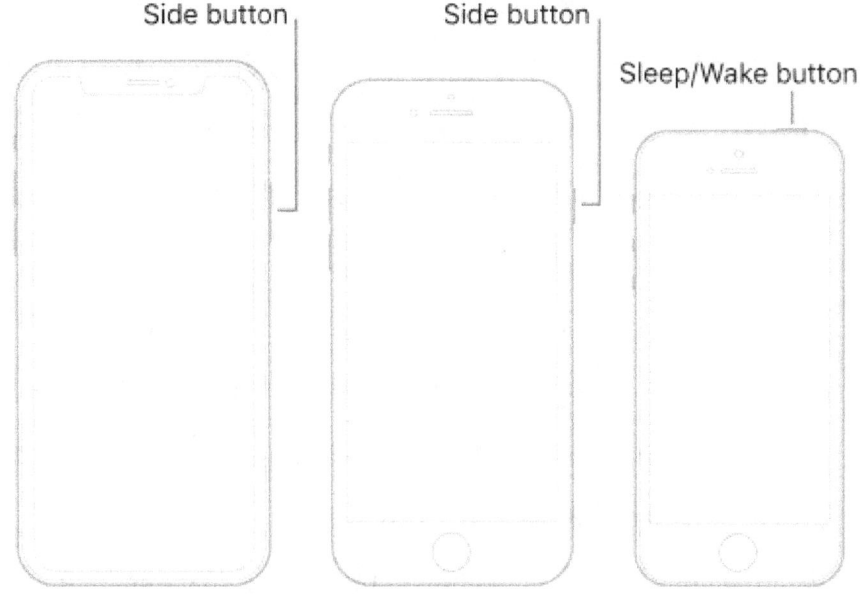

Touch Manual Setup, and then follow the instructions that appear on the screen.

- If you have another iPhone, iPad, or iPod touch with iOS 11, iPad 13, or later, you can use Quick Launch to automatically configure your new device. Move the two devices close together, and then follow the instructions that appear on the screen to securely copy preference, settings, and the iCloud keychain. You can then restore the rest of the data and contents of the new device from your iCloud backup.

- Or, if both devices have iOS 12.4, iPad 13, or later, you can transfer all data wirelessly from your previous device to your new device. Hold your devices close together and connect them to a power source until the migration process is complete.
- You can also transfer your data between devices via a wired connection. See Using Quick Launch to transfer data from your previous iOS device to your new iPhone, iPad, or iPod touch.
- If you're blind or visually impaired, press the side button (on an iPhone with a Face ID), then click the Home button three times (for other iPhone models) to turn on VoiceOver, a screen reader. You can double-click the screen with three fingers to turn on zoom.

Switch from Android device to iPhone

After setting up your new iPhone, you can automatically transfer your data from an Android device.

Note: You can only use Move to iOS when you first set up the iPhone. If you've already completed the installation and want to use the Move to the iOS option, you'll need to delete iPhone and restart it or move the data manually.

See the Apple Support article: Manually transfer content from your Android device to your iPhone, iPad, or iPod touch.

On your Android 4.0 or later device, read Apple's support article: Migrating from Android to iPhone, iPad, or iPod touch, and download the Migrate to iOS app.

On the iPhone, do the following:

Follow the installation assistant.

Tap Move Data from Android on the app and computer screen.

On your Android device, do the following:

Turn on Wi-Fi.

Open Move to iOS.

Follow the instructions on the screen.

To reduce the risk of personal injury, read the Important iPhone Safety Information before using iPhone.

See also

Connect your iPhone and computer via USB

Apple Support Article: Disable the activation lock

Apple Support Article: Help your child set up their iPhone, iPad, or iPod touch

Set up mobile service on iPhone
iPhone requires a SIM card from your carrier for your mobile connection. contact your operator to set up a mobile plan.

You can use iPhone nano-SIM to connect to an operator network. You can use nano-SIM and eSIM on an iPhone that supports a dual SIM card (not available in all countries or regions).

Here are some of the many ways to use Dual SIM:

Use one number for your business and another number for personal conversations.

Add a local data package if you are traveling to another region.

Have a separate voice and data package.

Note: To use two different carriers, your iPhone must be unlocked.

Install the nano-SIM card

Insert a paper clip or SIM release tool into the small slot on the SIM tray, and then slide it toward the iPhone to release the tray.

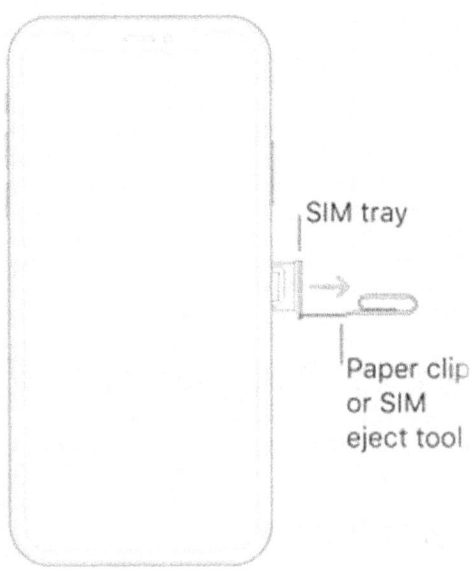

Insert a paper clip or SIM release tool into the small hole in the tray on the right side of the iPhone to release and remove the tray.

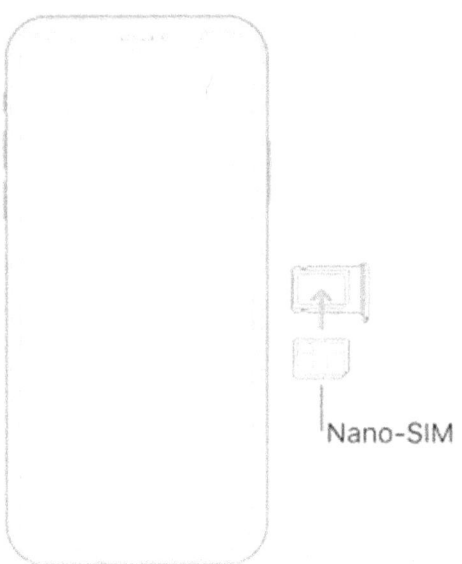

The shape and orientation of the SIM tray depend on your iPhone model and region.

Remove the account from iPhone.
- Insert the nano-SIM card into the tray. The sloping corner determines the correct direction.
- A nano-SIM is inserted in your iPhone account; the square corner is in the upper right corner.
- Put the tray back on the iPhone.
- If you have previously set a PIN on the nano-SIM card, you will be asked to enter the PIN carefully.
- Never try to guess your SIM PIN. Improper guessing can permanently lock the SIM card and

only call or use a mobile data service provider if you receive a new SIM card. See the Apple Support Article Using a SIM Card for iPhone or iPad.

- To connect to GSM networks and some CDMA networks, you may need a nano-SIM to use mobile services. iPhone enabled on a CDMA wireless network can also use a nano-SIM card to connect to the GSM network for international roaming. iPhone is subject to wireless carrier policies, which may include restrictions on switching and roaming, even if you have a required minimum service contract. Contact your wireless service provider for more information. The availability of mobile features depends on your wireless network, iPhone model, and your location.

Set up your mobile package using eSIM
On models that support eSIM, eSIM from your service provider is digitally stored on the iPhone.

Choose Settings> Mobile, then tap Add Mobile Package.

Position iPhone so that the provider's QR code appears in the frame, or enter the details manually. You may need to enter a verification code provided by your service provider.

Tap Add Mobile Package.

If the new plan is the second line, follow the instructions on the screen to specify how you want the plans to work.

Alternatively, you can activate your mobile through the operator's app (if supported). Go to the App Store, download the carrier app, and then use the app to activate a mobile subscription.

You can store more than one eSIM on iPhone, but you can only use one eSIM at a time. To switch to eSIM, go to Settings> Mobile, tap the plan you want to use, and then tap Turn on.

If you have a nano-SIM, you can use it as a second line. See Apple Support Article: Using Dual SIM with eSIM.

CHAPTER TWO

MANAGE YOUR MOBILE PLANS

When setting up models with dual SIM cards, you can choose how the iPhone uses each line. To change the settings later, do the following:

Select Settings> Mobile.

Do the following:

Tap Mobile data, then select a default row. To allow the iPhone to use both lines, turn on Enable Mobile Data Exchange, depending on coverage and availability.

Roaming charges may apply if data roaming is enabled and you are outside your home network.

Touch Default tone, then selects a line.

Tap a line under mobile packages and change settings like mobile planet tag, Wi-Fi calls (if available from your service provider), calls to other devices, or SIM PIN. The label is shown in Phone, Messages, and Contacts.

When using a dual SIM, Do The following:
- Wi-Fi calls must be turned on on one line so that that line can receive calls while the other line is in use during a call. If you receive a call on one line while you're on the other and you don't have a Wi-Fi connection, iPhone uses the mobile data on the line you're using to make the call on the other line. Costs may apply. The line used for the call must be enabled for data traffic in your mobile data settings (either with mobile data enabled as default or non-default line) to answer the call on the other line.
- If you do not turn on Wi-Fi calls on a line, all calls on that line (including emergency calls) will go directly to the answering machine (if available from your service provider) when the other line is in use; you will not be notified of missed calls.
- If you set up conditional redirection if available from your service provider from one line to another.

CHAPTER THREE

CONNECT THE IPHONE TO THE INTERNET

Connect the iPhone to the Internet using an available Wi-Fi or mobile network.

Connect iPhone to a Wi-Fi network

Select Settings> Wi-Fi, then turn on Wi-Fi.

Tap one of the following:

Network: Enter the password if required.

Other: Connects to a hidden network. Enter the hidden network name, security type, and password.

If the Wi-Fi icon appears at the top of the screen, the iPhone connects to a Wi-Fi network. (To confirm this, you can view the webpage by opening safari)iPhone reconnects when you return to the same location.

Join a personal hotspot
If an iPad (Wi-Fi + mobile) or another iPhone shares a personal hotspot, you can use a mobile Internet connection.

Select Settings> Wi-Fi, then select the name of the device that shares your hotspot.

If you're prompted for a password on iPhone, enter the password in Settings> Mobile> Personal Hotspot on the personal hotspot sharing device.

Connect iPhone to a mobile network

iPhone automatically connects to your carrier's mobile data network if a Wi-Fi network isn't available. If the iPhone isn't connected, check the following:

Make sure the SIM card is turned on and unlocked. See Setting Up a Mobile Service on iPhone.

Select Settings> Mobile.

Make sure mobile data is turned on. For dual SIM models, tap Mobile data, then confirm the selected line. (You can only select one line for mobile data.)

When you need an Internet connection, the iPhone does the following in order, until the connection is established:

You are trying to connect to the most recently used available Wi-Fi network

Displays a list of Wi-Fi networks within range and connects to the selected network

Connects to the operator's mobile data network

If you do not have a Wi-Fi connection to the Internet, applications and services may transfer data over the operator's mobile network, which may incur additional charges. Contact your service provider for information about mobile data plan prices. To manage mobile data usage, see View or change mobile settings on iPhone.

CHAPTER FOUR

MANAGE APPLE ID AND ICLOUD SETTINGS ON IPHONE

An Apple ID is the account you use to access Apple services such as the App Store, iTunes Store, Apple Books, Apple Music, FaceTime, iCloud, iMessage, and more.

With iCloud, you can store photos, videos, documents, music, apps, and more securely — and keep them up to date on all your devices. With iCloud, you can easily share your photos, calendars, locations, and more with your friends and family. You can even use iCloud to locate your iPhone if you lose it.

iCloud provides a free email account and 5GB of storage for emails, documents, photos and videos, and backups. Purchased music, apps, TV shows, and books are not included in the available storage. You can update your iCloud storage directly from the iPhone.

Note: Some iCloud features have minimum system requirements. The availability of iCloud and its services varies by country and region.

Sign in with your Apple ID

If you did not log in during configuration, do the following:

- Go to settings.
- Tap Sign in to iPhone.
- Enter your Apple ID and password.
- If you do not have an Apple ID, you can create one.
- To protect your account with 2-factor authentication, enter the six-digit verification code.
- If you forget your Apple ID or password, see Recovering Your Apple ID Website.
- Change Apple ID settings
- Select Settings> [your name].
- Do one of the following:
- Update your contact information
- Change password
- Managing family sharing

Change iCloud settings

Go to Settings> [your name]> iCloud.

The iCloud settings screen, which displays the iCloud storage meter and a list of applications and features you can use using iCloud, including Mail, Contacts, and Messaging.

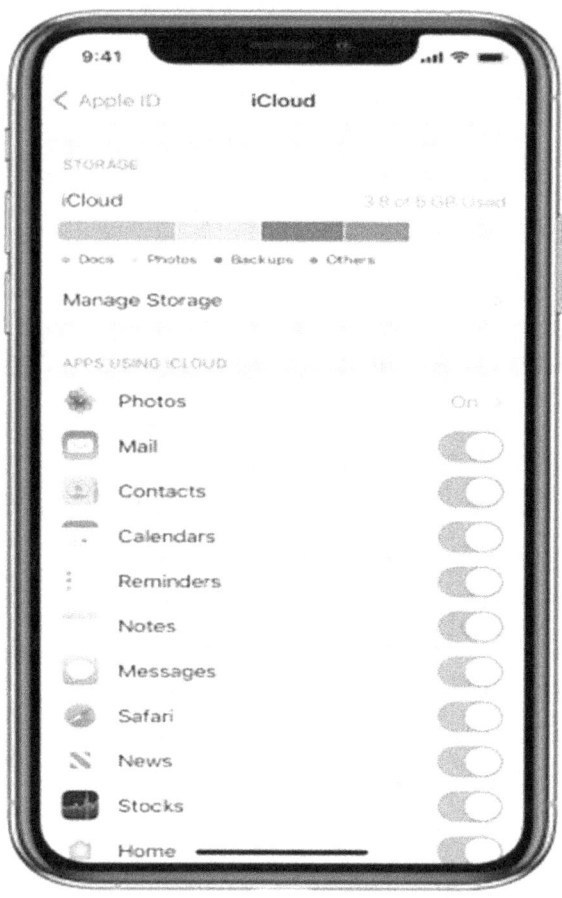

Do one of the following:

View the status of your iCloud storage.

Update your iCloud storage - tap Manage storage> Change storage schedule.

Turn on features you want to use, such as photos, emails, contacts, and messages.

Ways to use iCloud on iPhone
Keep the following content up to date:

- Messages, emails, contacts, calendars, notes, and reminders
- Photos and videos; see Using iCloud Photos on iPhone
- Music, apps, and books
- Documents; see Setting Up iCloud Drive on iPhone
- Bookmarks, reading list, and websites opened in Safari; see Browsing the Internet Using iPhone Safari
- Passwords and credit cards; see Making Passwords Available on All Devices with the iPhone and iCloud Keychain
- You can also do the following:
- View iCloud data on iPhone, iPad, iPod touch, Apple Watch, Mac, and iCloud.com (Mac or Windows PC).

- Share your photos and videos with the people you choose. See Sharing iPhone photos with shared albums in iCloud.
- Share your iCloud storage with plans of 200 GB or more with up to five other family members.

- Find a missing iPhone, iPad, iPod touch, Apple Watch, Mac, or AirPod that you or your family own. See Finding a Device in Finding iPhone.
- Find your friends and family; You, your friends and family can share places, follow each other, and see everyone's location on a map. See Finding a Friend on iPhone.
- Back up and restore your data. See Backing Up Your
- iPhone.

Wake up and unlock iPhone

iPhone turns off the screen to save power, locks it for safety, and sleeps when not in use. You can quickly wake up and unlock your iPhone when you want to use it again.

CHAPTER FIVE

WAKE UP IPHONE

one of the following should be done to wake your iPhone:

look for were the side button is located and press the side button or the Sleep / Wake button

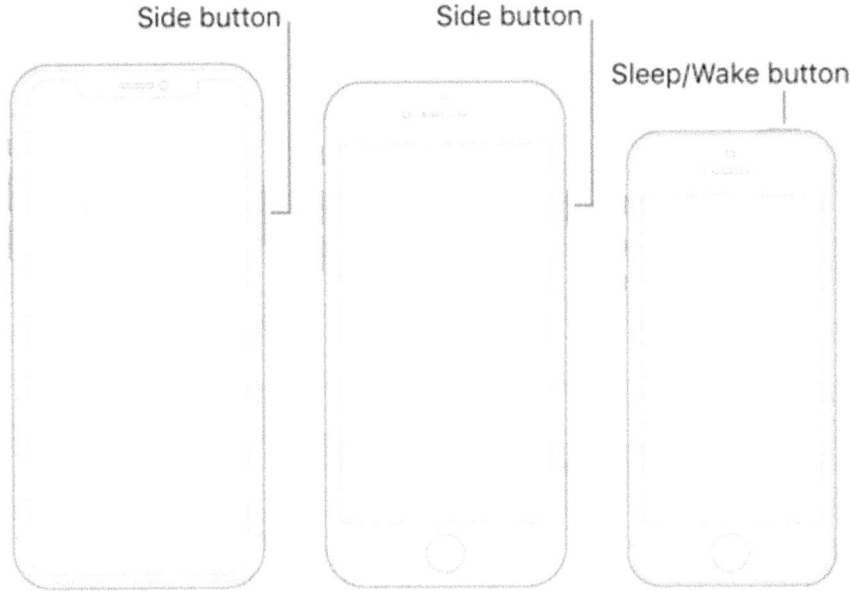

- Lift the iPhone. To set the alarm to go off, you can turn it off in Settings> Display & Brightness.

- Illustration showing how to wake up your iPhone.
- Touch the screen (supported models).
- An illustration that taps the screen to wake up the iPhone.

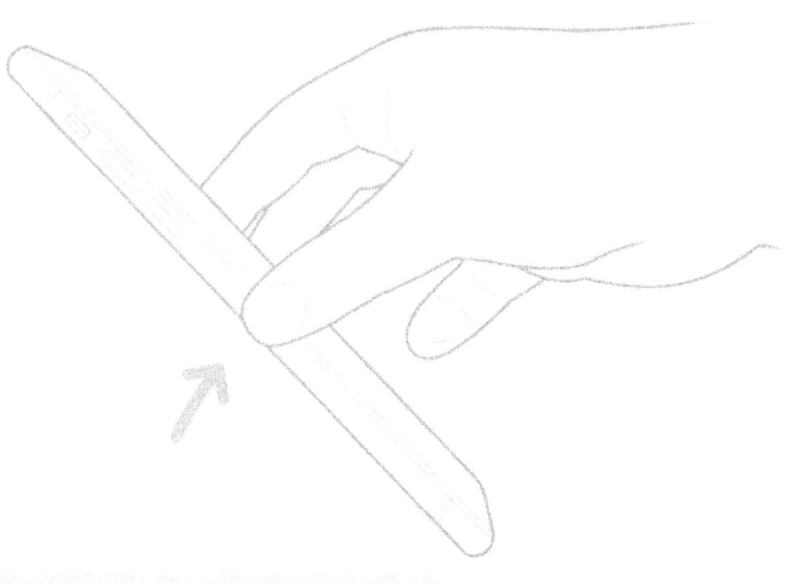

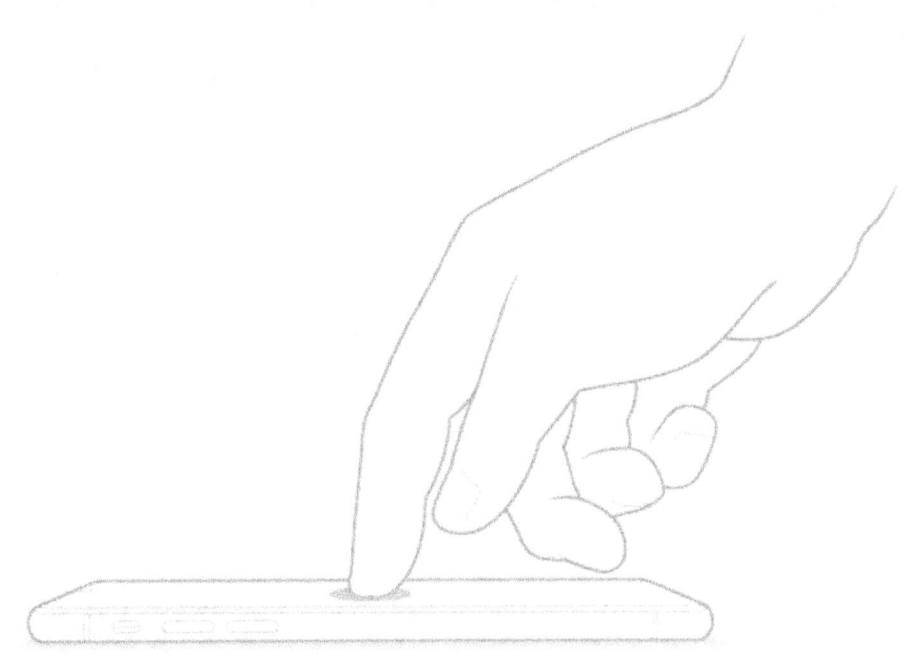

Unlock iPhone with Face ID
If you didn't configure Face ID when setting up iPhone, see Configuring Face ID on iPhone.

On supported models, tap the screen or lift the iPhone to wake it up, then watch the iPhone.

The lock icon is animated from closed to open to indicate that iPhone is unlocked.

Drag the bottom of the screen up.

To lock iPhone again, press the side key. If you don't touch the screen for a minute it automatically locks the screen.

Unlock iPhone with Touch ID
If you didn't configure Touch ID when setting up iPhone, see Configuring Touch ID on iPhone.

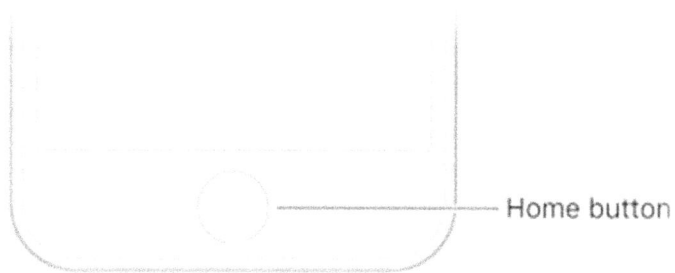

On an iPhone with a Home button, press the Home button with your finger registered using Touch ID.

The home button on the bottom of the iPhone.To lock iPhone again, press the side button or the Sleep / Wake button (depending on the model). if you don't touch the screen for a minute the screen automatically locks

Unlock iPhone with a password
If you didn't create a password when you set up the iPhone, see Entering a Password on iPhone.

Twirl up from the bottom of the lock screen (on iPhone with Face ID) or press Home (for other iPhone models).

Enter the password (if you set the iPhone to require a password).

To lock iPhone again, press the side button or the Sleep / Wake button (depending on the model). if you don't touch the screen for a minute its screen will automatically lock.

CHAPTER SIX

LEARN THE BASIC GESTURES OF INTERACTING WITH YOUR IPHONE

You can control your iPhone and apps with a few simple gestures - tap, tap and hold, slide, scroll, and zoom.

Symbol	Gesture

Icon symbolizing the movement of pressure.

Tap your finger on the screen.

Touch and hold icon symbolizing movement.

Press and hold. Touch and hold an application or Control Center to preview content and perform quick actions. From the Home screen or the App Library, tap and hold the app icon to open the Quick Actions menu.

An icon that symbolizes swipe motion.

Swipe. Move your finger quickly on the screen.

Icon symbolizing rolling motion.

Blue. Move your finger on the screen without lifting it. For example, in Photos, you can see more by dragging a list up or down. Drag your finger to scroll quickly; tap the screen to stop scrolling.

Icon symbolizing the zoom movement.

Zoom. Close your two fingers together and place them on the screen. Drag them apart to zoom in or move toward each other to zoom out.

To zoom in on a webpage or an image double click, and do the same thing to zoom out...

To view and hold the map you will have to double click, thereafter you drag up or down to zoom in or out.

Adjust the iPhone volume
While on the phone making calls or listening to songs or other media, the buttons on the side of the iPhone adjust

the volume. Otherwise, the keys control the volume of the ringtone, alarms, and other sound effects. You can also use Siri to increase or decrease the volume.

Ask Siri. Switch up the volume or Turn down the volume. Learn how to ask Siri.

For important information on preventing hearing damage, see Important iPhone Safety Information.

Lock the ring volume and alarm volume in Settings
Go to Settings.

Tap Sounds and Haptics (supported models) or Sounds (other iPhone models).

Turn off Change with.

Adjust the volume in the control center
If the iPhone is locked or you are using an app, you can adjust the volume in Control Center.

Open the Control Center, then drag the volume slider.

Limits the headphone volume
You can choose to limit the maximum headphone volume for music and videos.

Go to Settings.

Tap Sounds and Haptics (supported models) or Sounds (other iPhone models).

Press Mute, turn on Mute and then drag the slider to select the maximum decibel level for the headphone sound.

The Mute Loud screen, which is a button to turn the sound on or off, displays a slider to change the maximum decibel level and the selected limit of 85 decibels.

If you have turned on the screen time in Settings, you can prevent the volume of the headphones from changing. Select Settings> Screen time> Content and privacy restrictions> Mute, and select Disable.

Temporarily mute calls, alerts, and alerts

Open the control center and press Do not disturb. See Do Not Disturb iPhone.

Set iPhone to call or silent mode
- To set iPhone to ring mode, turn off the ring / mute switch.

- Top of the front of the iPhone with the display pointing to the Ring / Silent switch.
- In-ring mode, the iPhone plays all sounds. In silent mode the orange switch, the iPhone does not ring or play alarms or other sound effects (but the iPhone can still vibrate).
- Alerts, audio apps like Music, and many games play sounds through the built-in speaker, even when the iPhone is in silent mode. In some regions, the sound effects of the camera, voice memos, and emergency

alerts can be played back even if the Ring / Mute switch is muted.

CHAPTER SEVEN

SWAP THE SOUNDS AND VIBRATIONS OF THE IPHONE

In Settings, you can change iPhone sounds when you receive a call, SMS, voicemail, email, reminder, or other types of notification.

On supported models, you will feel a touch - called haptic feedback - after performing certain actions, such as when you press and hold the camera icon on the Home screen.

Adjust the sound and vibration options
move to Settings> Sounds & Happiness (for supported models) or Sounds (for other iPhone models).

To adjust the volume for all sounds, drag the slider under Ringtones and Alerts

To set sounds and vibration patterns for sounds, tap a sound type, such as a ringtone or text sound.

Do one of the following:

Select a sound (scroll to see all).

Ringtones are used to play incoming calls, time alarms, and a timer; text tones are used for text messages, new voicemail, and other notifications.

Tap Vibrate, then select a vibration pattern or tap Create new vibration to create your own.

Turn haptic feedback on or off
go to Settings> Sounds & Haptics.

Turn System Haptics on or off.

When System Haptics is turned off, you will not hear or feel vibrations from incoming calls and alerts.

Tip: If you do not hear or see incoming calls and alerts while waiting for them, open the Control Center and make sure Do Not Disturb is turned on. If the Do Not Disturb button is highlighted, press it to turn off Do Not Disturb. (If the Do Not Disturb feature is turned on, the Do Not Disturb icon also appears in the status bar.)

CHAPTER EIGHT

FIND IPHONE SETTINGS

you can find iPhone settings you want to change, such as your password, alert sounds, and more.

- From the Home screen (or in the App Library), tap Settings.

Tap Settings to change your iPhone settings (volume, screen brightness, and more).

- Swipe down to display the search box, type a phrase, such as "iCloud", then tap an option

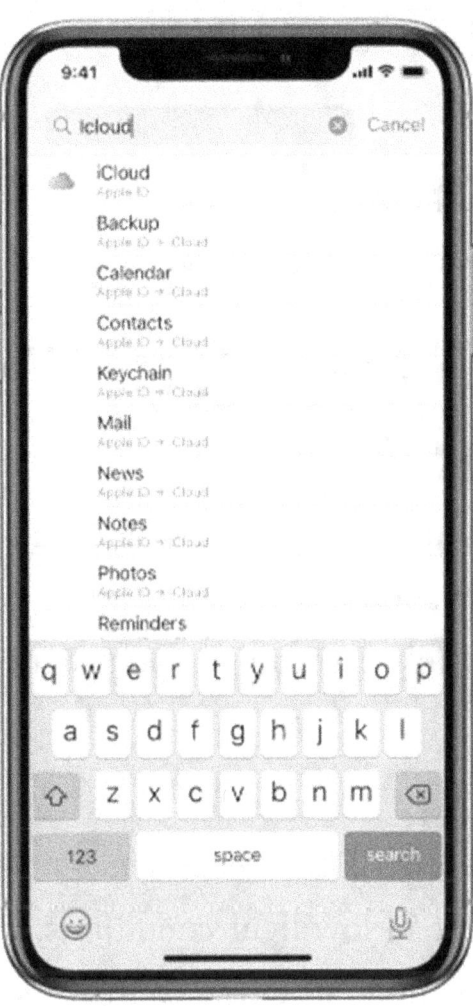

CHAPTER NINE

ARRANGE THE BRIGHTNESS AND COLOR OF THE IPHONE SCREEN

On an iPhone, you can dim the screen to extend battery life, use the Night Shift function, set the dark mode, and adjust the screen automatically according to your lighting conditions.

Turn dark mode on or off
The dark mode gives the entire iPhone experience a dark color scheme that is perfect in low light conditions. You can choose to turn on dark mode from the control center or set it to turn on automatically at night (or on a custom schedule) in Settings. When Dark Mode is on, for example, you can use the iPhone while reading in bed without disturbing the person next to you.

Do one of the following: Open the Control Center, press and hold the Brightness button, and then press the Appearance button to turn dark mode on or off.

Select Settings> Display and brightness, select Dark to turn on dark mode or Light to turn it off.

Set dark mode to turn on and off automatically
Select Settings> Display and brightness.

Turn on Automatic, and then tap Options.

Select Sunset to Sunrise Or Custom Schedule.

If you select a Custom plan, press the plan to plan when dark mode is on or off.

If you select Sunset and Sunrise, iPhone uses your clock and geographic location to determine when it's night.

Adjust the screen brightness manually
To make the iPhone screen brighter or brighter, do one of the following:

Open the control center and drag the brightness button.

Select Settings> Display and brightness, and then drag the slider.

Adjust the screen brightness automatically

The iPhone uses the built-in ambient light sensor to adjust the screen brightness for current lighting conditions.

Go to Settings> Availability.

Touch Screen and Text Size, then turn on Auto-Brightness.

Turn True Tone on or off
On supported models, turn on the True Tone feature to automatically adjust the color and intensity of the screen to the ambient light.

- Do one of the following:
- Open the Control Center, press and hold the brightness button, and then press the True Tone button to turn True Tone on or off.
- Go to Settings> Display and brightness, then turn the real sound on or off.
- Turn Night Shift on or off
- You can also turn on Night Shift manually, which is useful if you are in a dark room during the day.
- Open the Control Center, press and hold Brightness, and then press Night Shift.
- Set night shift for automatic on and off
- Night Shift allows you to move the screen colors to the warmer end of the spectrum at night and make it easier to see the screen in your eyes.
- Select Settings> Display and brightness> Nightshift.

- Turn on Scheduled.
- To adjust the Night Shift color balance, drag the slider below the color temperature toward the warmer or cooler end of the spectrum.

Tap Off, then select Sunset to Sunrise or Custom Plan.

- If you select Custom Schedule, press the options to schedule Night Shift on and off.
- If you select Sunset and Sunrise, iPhone uses your clock and geographic location to determine when it's night.
- Note: Sunset to Sunrise is not available if you have turned off Location Services in Settings> Privacy, or if you have turned off the time zone setting in Settings> Privacy> Location Services> System Services.

CHAPTER TEN

ENLARGE IPHONE SCREEN WITH SCREEN ZOOM

- You can see controls that appear on a larger screen on an iPhone that has a view zoom.
- Select Settings> Display and brightness.
- Press View (during zoom).
- Select Magnified, then press Set.
- See models that support screen magnification.
- For more zoom features, see Zoom In on the iPhone screen.

Change the iPhone name

You can change the iPhone name used by iCloud, AirDrop, personal hotspot, and computer.

Select Settings> General> About> Name.

Tap Clear text, enter a new name and then tap Done.

Adjust the date and time on the iPhone

By default, the date and time on the lock screen are set automatically based on your location. If they are incorrect, you can adjust them.

Select Settings> General> Date & time.

Turn on one of the following:

Automatic setup: iPhone gets the correct time over the network and updates it to the time zone you are in. certain networks do not support network time, in some countries or regions iPhone may not automatically set local time.

24-hour time: (Not available in all countries or regions) iPhone displays hours from 0 to 23 hours.

In other To adjust the default date and time, switch off the Auto setting, and then change the displayed date and time.

Enter language and region on iPhone

You can specify the iPhone language and region during installation. If you are traveling or moving, you can change the language or region.

- Select Settings> General> Language & region.
- Set the following:
- The language of the iPhone
- The regions
- The calendar format
- Temperature unit (Celsius or Fahrenheit)
- If you choose To add another language and keyboard to iPhone, click Add language, then select a language.
- You can add or change keyboards on iPhone.

Set up email, contacts, and calendar accounts

In addition to the apps that come with the iPhone app and are used with the iCloud app, the iPhone also works with

Microsoft Exchange and the most popular Internet records, contacts, and calendar services. You can choose to create accounts for these services.

Create an email account

- Select Settings> Email> Accounts> Add account.
- Do one of the following: Tap an email service, such as iCloud or Microsoft Exchange, and enter your email for information. Tap More, tap Add Email Account, and then enter your information to create a new account.
- Create a contact account
- Go to Settings > Contacts > Accounts> Add account> Other Touch Add LDAP Account or Add CardDAV Account (if supported by your organization), and then enter the server and account information. See Using other contacts on iPhone.

Create a calendar account
- Select Settings> Calendar> Accounts> Add Account.
- Touch Other, and then do one of the following:
- If you would choose to add a calendar account: Tap Add CalDAV Account, then enter the server and

account information. see Setting Up Multiple Calendars on iPhone.

- To subscribe to iCal (.ics) calendars: Tap Add subscription calendar, then enter the URL of the .ics file you want to subscribe to; or import a .ics file from Mail.

-

Access features on the iPhone lock screen

When you turn on or wake up iPhone, a lock screen appears showing the current time and date and the latest alerts.open the Camera and Control Center, get information from your favorite taps in an instant, and more. On the lock screen, you can view alerts.

Access features and information with Screen Lock

You can quickly access useful features and information from the lock screen, even when the iPhone is locked. On the lock screen, do one of the following:

To open the camera: Slide your finger to the left. On supported models, press and hold the camera key, then lift your finger. (See Taking pictures with your iPhone camera.)

Open Control Center: Slide your finger down from the top right corner (on iPhones with Face ID) or slide up from the bottom of the screen (on other iPhone models). (See Using and Customizing the Control Center on iPhone.)

To view previous alerts: Swipe up from the center. (See View and reply to alerts on iPhone.)

See today's view: swipe to the right. (See Adding Gadgets to the iPhone Home Screen.)

To select information available from the lock screen, see Checking Access to Information on the iPhone Lock Screen.
- View screen lock alerts
- Go to Settings> Notifications.
- Tap Show preview, then tap Always. Alerts preview includes text messages, email codes, and calendar invitation details. See View and respond to alerts on the iPhone.

Open apps on iPhone

Get to know the home screen and apps on the iPhone. On the home screen, all applications are arranged in pages. If you need more applications, more pages will be added.

You can also find your apps in the App Library - somewhere at the bottom of the home screen pages where your apps are organized in a simple, easy-to-navigate view.

Open applications on the home screen
To jump to the home screen, slide your finger up from the bottom edge of the screen (on an iPhone with Face ID) or press the Home button (on an iPhone with a Home button).

Swipe left or right to browse apps on other pages on the Home screen.

- Illustration showing your finger to scroll through other pages on the Home screen.
- If you open an app, click the icon.
- To return to the first home screen, slide your finger up from the bottom edge of the screen (on an

iPhone with Face ID) or press the Home button (on an iPhone with a Home button).

- You can move, organize, or remove applications. See Moving and Organizing Apps on iPhone and Removing Apps from iPhone.

-

Explore the app library

The app library automatically sorts apps into categories such as creativity, social, entertainment, and more. The most used apps are near the top of the screen and at the top of the tabs, so you can easily find and open them.

To find the application library, go to the home screen and swipe left for all home screen pages.

The iPhone app library organizes apps by category (tools, creativity, social, entertainment, etc.)

- To open an app: Tap the app if it's visible.

- Expand category: If there are several apps in a category below the top-level (indicated by some small app icons), tap the small icons to see all the apps in the category.
- Search for apps: Tap the search box at the top of the screen to list apps alphabetically. Enter the name of the application in the search box to find it.
- Perform quick actions: Tap and hold an app to open the quick action menu.
- Note: If it takes too long and you hold down an app before selecting a quick action, all apps will start to stutter. Tap Done (on an iPhone with Face ID) or tap the Home button (on an iPhone with a Home button) and try again.
- Add an app to the Home screen: Tap and hold an app to open the Quick Actions menu, then select Add to the Home screen (available only if the app is not already on the Home screen). The app is still displayed in the app library.
- To delete an app from iPhone: Tap and hold the app, select Delete app, and then tap Delete. (This removes the application from the Home screen and

the application directory.) See Uninstalling Applications.

You can add new apps from the App Store to the Home screen and the App Library, or just the App Library. From the Home screen, tap Settings> Home screen, and then tap Add. You can also make app alert tags appear in App Library apps by turning on View in App Library.

Hide and show pages on the home screen
Since you can find all the apps in the App Library, you may not need as many home screen pages for the apps. You can hide some Home screen pages, and bring the App Library closer to the first page of the Home screen. (If you want to see the pages again, you can view them.)

- Tap and hold any app on the Home screen, then tap Edit Home screen.
- Apps are starting to giggle.
- Tap the dots at the bottom of the screen.
- Thumbnails of the home screen pages are checked below.
- To hide pages, tap to clear the thumbnail of unwanted pages.
- To show hidden pages, press to add checkmarks.

- Double-click the Done button (on an iPhone with Face ID), or double-tap the Home button (on an iPhone with a Home button).
- By hiding the extra home screen pages, you can move to the application directory (and back) from the first page of the home screen with just one or two dragons.
- Note: If the pages on the Home screen are hidden, new apps downloaded from the App Store can be added to the app directory instead of the Home screen.

Take a screenshot or screenshot on iPhone

record an action on the screen to share with others or use in documents You can take a picture of the screen as it appears.

Take a screenshot
Do one of the following:

On an iPhone with Face ID: Press and release the side and volume up buttons simultaneously.

For iPhones with the Home button: Press and release the Home button and the side button or the Sleep / Wake button at the same time (depending on the model).

Tap the screen in the lower-left corner, then tap Done.

Select Save To Photos, Save To Files, or Delete Screenshot.

If you select Save to Photos, you can see it in the Screenshot album in the Photos app or the All Photos album if iCloud Photos is turned on in Settings> Photos

To create a PDF of a webpage, document, or email, take a screenshot, tap the thumbnail, and then tap Entire Page.

Create a screenshot
You can create a screenshot and record audio on the iPhone.

Go to Settings> Control Center, then tap Paste Screenshot next to Screenshot.

Open the Control Center, tap Screenshot, and then wait for the countdown to three seconds.

open the Control Center, tap Selected screen or the red status bar at the top of the screen, then tap Stop To stop recording.

Open images and select a screenshot.

Change or lock the screen orientation on iPhone

Many apps give a different view when you rotate the iPhone.

In the background, the iPhone displays a calendar screen that shows the events of a day in a vertical direction. in the foreground, the iPhone is placed in a landscape position that shows calendar events throughout the week that is the same day.

Lock or unlock screen orientation

You can lock the screen orientation so that it does not change when you rotate the iPhone.

Open Control Center, and then tap Lock Direction.

If the screen orientation is locked, the Orientation Lock icon appears in the status bar (on supported models).

Change iPhone wallpaper

On the iPhone, select a photo or image as the background for the lock screen or the home screen. You can choose between dynamic and still images.

With the background settings screen, a new background selection button at the top, and images of the screen lock and home screen with the current background.

Change background

Select Settings> Wallpaper> Select New Wallpaper.

Do one of the following:

Select a preset image from the group at the top of the screen (Dynamic, Snapshots, etc.).

The background marked with the Appearance button changes its appearance when dark mode is on.

Select your photo (tap the album, then tap the photo).

To place the selected image, pinch it to zoom in, then use your finger to move the image. Tighten the clip to zoom out.

Press the Parallax Effect button to turn on Perspective Zoom (available with some background selections), which moves the background as if you changed the viewpoint.

The perspective zoom option does not appear when motion reduction is turned on (in Settings> Accessibility>

Motion). stop the movement of screen elements on the iPhone.

Press Setup, and then select one of the following:

Set the screen lock

Set the home screen

Set up both

To turn on Perspective Zoom for already set wallpapers, go to Settings> Wallpaper, tap Lock screen or Home screen, and then tap Perspective zoom.

Set a Live Photo as the background for the lock screen
To set Live Photo as wallpaper, press and hold the lock screen to play Live Photo - on all iPhone models except iPhone SE (Generation 1).

Select Settings> Wallpaper> Select New Wallpaper.

Do one of the following:

Tap Live, then select Live Photo.

Tap the Live Photos album, then select a Live photo (you may have to wait for the download).

Tap Layout, and then select Display Layout or Both.

Switch between iPhone apps

Open App Switcher to quickly switch from one app to another on iPhone. When you switch back, you can continue where you left off.

The app switch. Icons for open applications are displayed at the top, and the current screen for each application is displayed below the icon.

Use Application Switcher

To see all open applications in the Application Switcher, do one of the following:

On an iPhone with face ID: Slide your finger up from the bottom edge and pause in the center of the screen.

For iPhones with the Home button: Double-click the Home button.

To browse open apps, swipe right, then tap the app you want to use.

Switch between open applications
To quickly switch between open apps on iPhone with Face ID, slide the bottom edge of the screen left or right.

Multitasking picture in picture on iPhone

With Picture in Picture, you can use FaceTime or watch videos while using other applications.

A screen that shows the FaceTime conversation while you look at the Calendar app, which fills the rest of the screen.

If you are using FaceTime or watching a video, tap Picture in picture.

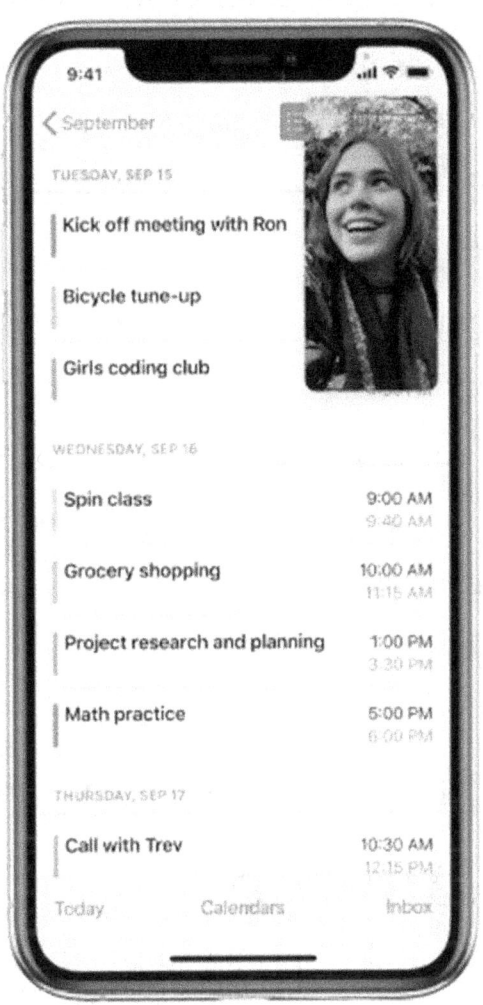

The video window shrinks to the corner of the screen so you can see the home screen and open other applications. In the video window that appears, you can do any of the following:

- Resize the video window: Pinch to enlarge the small video window. To shrink again, squeeze it.
- To show or hide controls: Tap the video window.
- To move the video window: Drag to another corner of the screen.
- To hide the video window: Swipe from the left or right edge of the screen.
- To close the video window: Press Close.
- To return to the entire FaceTime or video screen: Press the Full-Screen button in the small video window.

Type using the on-screen keyboard on iPhone

In iPhone apps, you can type and edit the text using the on-screen keyboard. You can also use an external keyboard and dictation to enter text.

Enter text using the on-screen keyboard
In any program that allows you to edit text, open the on-screen keyboard by tapping a text box. Press each key to type, or use QuickPath to type the word by sliding it from

one letter to another without lifting your finger (not available in all languages). Raise your finger to complete the word. You can use both methods while writing, and you can even switch in the middle of a sentence. (If you press Delete after typing a word, the entire word will be deleted.)

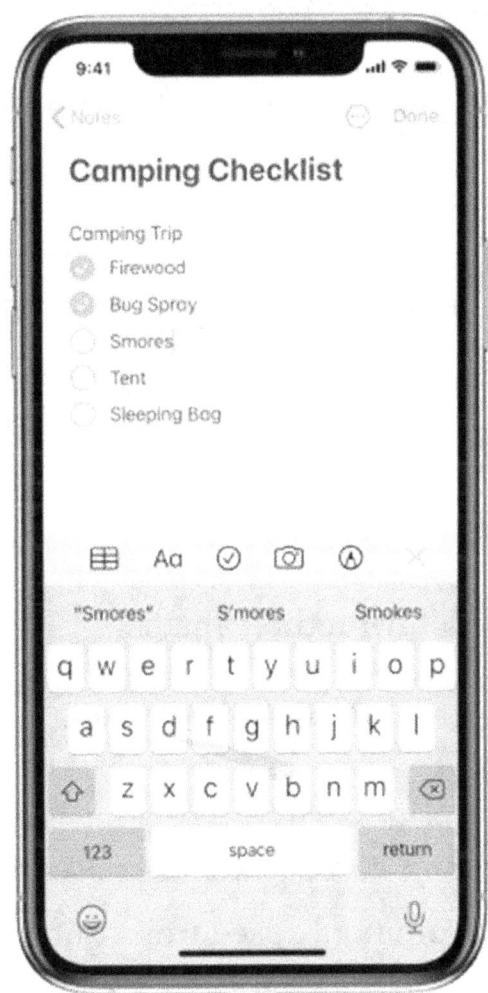

As you type, you will see suggested alternatives to the word you entered instead of predicting the next word.

When entering text, you can do any of the following:

To enter capital letters: Press Shift, or press Shift and slide to a letter.

To turn on Caps Lock: Double-click the Shift key.

End a sentence quickly with a period and space: Double-click on a space.

Correct spelling: Tap the misspelled word (underlined in red) to see the suggested corrections, then tap the suggestion to replace the word or enter the correction.

To enter numbers, punctuation, or symbols: Touch Numbers or Symbols.

To undo the last edit: Swipe left with three fingers.

To repeat the last edit: Swipe right with three fingers.

To enter an emoji: Touch Next Keyboard, Emoji, or Next Keyboard to switch to the emoji keyboard. You can search for an emoticon by typing a commonly used word, such as "heart" or "smiling face", in the search box above the emoji keyboard, and then scrolling through the emoticon that appears.

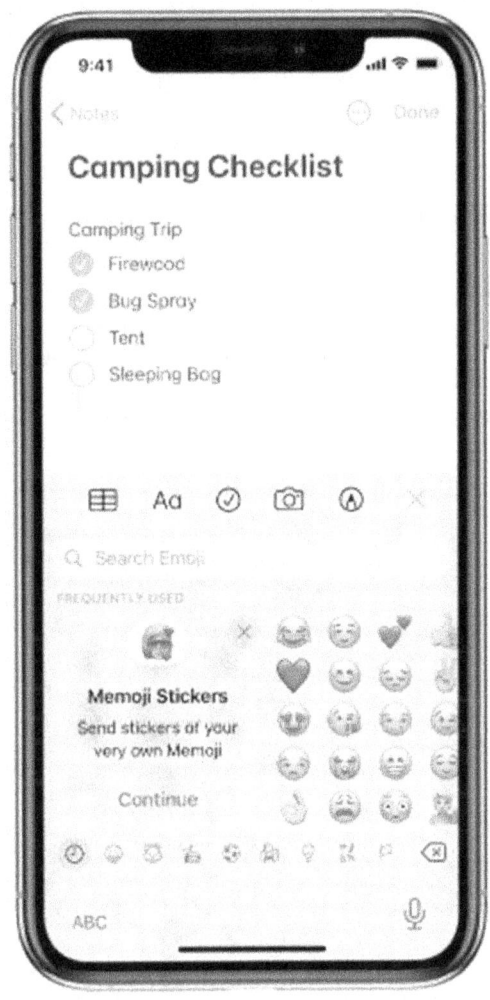

You can also dictate text, or enter text using the Magic Keyboard (sold separately).

Make the on-screen keyboard a trackpad

- Touch and hold the space bar with your finger until the keyboard turns light gray.
- Move the insertion point by swiping around the keyboard.
- To select text with a handle, keep touching and holding the keyboard until the handles appear at the insertion point, and move your fingers.
- To move the insertion point using the button, drag the insertion point to a new location before the handles appear.

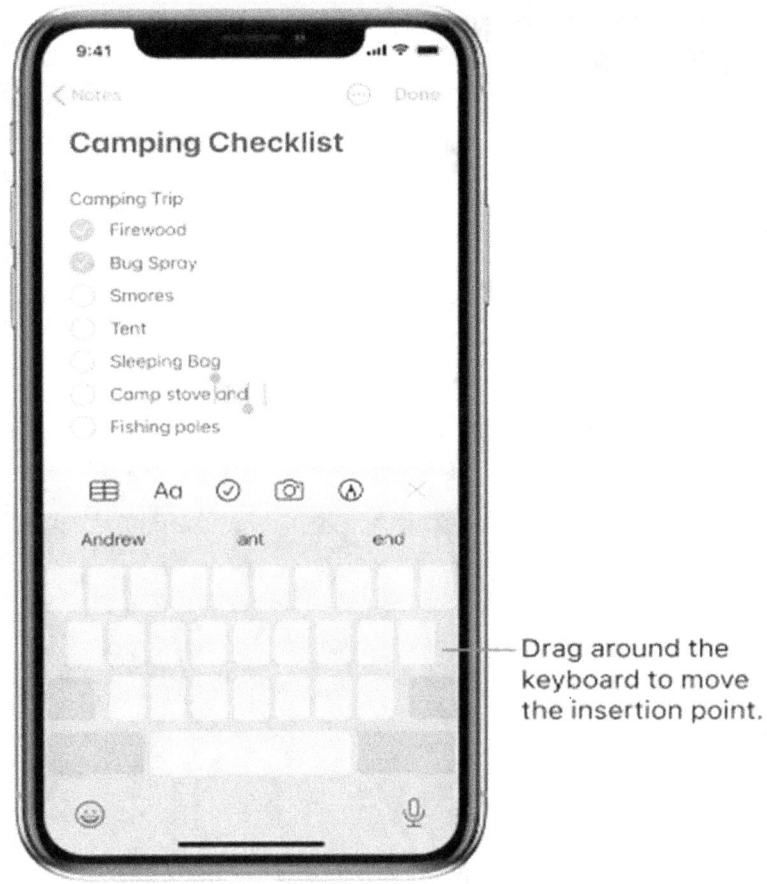

Drag around the keyboard to move the insertion point.

Enter letters with accents or other characters as you type

As you type, hold down the letter, number, or symbol associated with the desired character on the keyboard.

For example, to enter é, press, and hold e, slide to select the desired version.

You can do one of the following:

- On the Thai keyboard: Press and hold the corresponding Arabic numeral to select native numbers.
- On a Chinese, Japanese, or Arabic keyboard: Tap the suggested character or candidate you want to

type on the keyboard at the top of the keyboard, or swipe left to see more candidates.

- To see the full list of candidates, press the up arrow to the right. Press the down arrow to return to the list.

Move text
In the text editor, select the text you want to move.

Touch and hold the selected text until it appears, then drag it to another location in the app.

When you drag to the bottom or top of a long document, the document scrolls automatically.

A selected concept appears to rise as a result of the user pressing and holding the selection.

If you change your mind to move the text, lift your finger, or swipe the text off the screen before swiping.

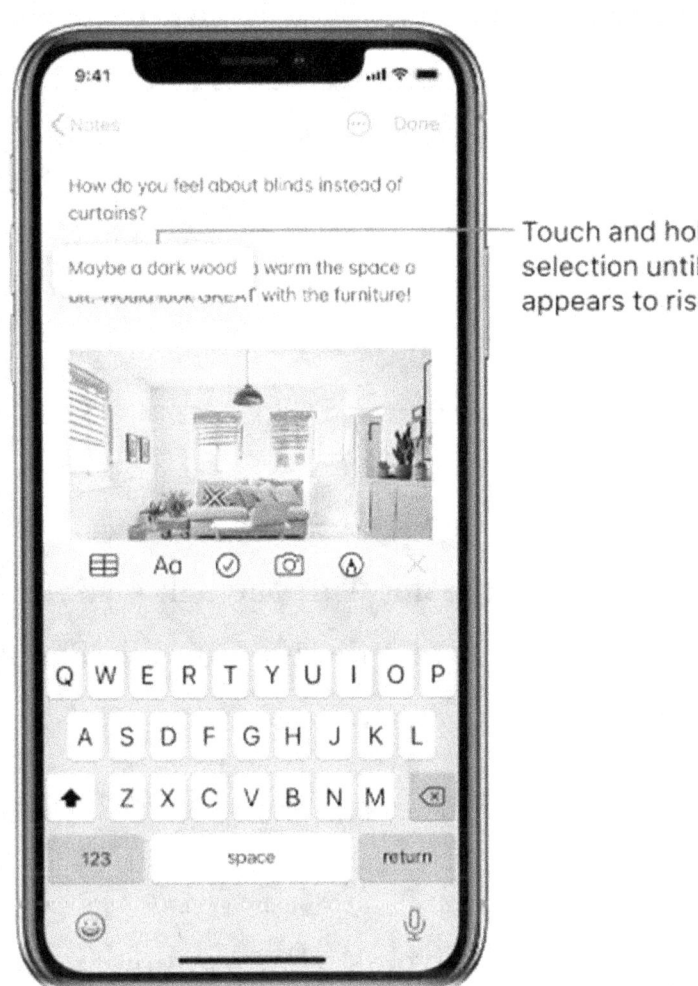

Specify write settings

You can turn on advanced typing features, such as predictive text and auto-correction, to help you type on iPhone.

While typing text on the on-screen keyboard, press and hold Next keyboard emoji or Change keyboard, and then tap Keyboard settings. Also go to Settings> General> Keyboard.

Switch typing functions on or off in the list (under All keyboards).

Enter with one hand
If you want to make typing easier with one hand, you can place the buttons closer to your thumb - on all iPhone models except iPhone SE (1st generation).

Touch and hold Next Keyboard Emoji or Replace Keyboard.

Touch one of the keyboard layouts. (For example, select the right layout button to move the keyboard to the right of the screen.)

To re-center, the keyboard, press the right or left edge of the keyboard.

CHAPTER ELEVEN

USE AIRDROP

AirDrop lets you share and receive photos, documents, and more with other nearby Apple devices.

Before you begin
Make sure the person you are sending is nearby, within range of Bluetooth and Wi-Fi.

Make sure you and the person you are sending are turned on via Wi-Fi and Bluetooth. If Personal Hotspot is turned on for any of you, turn it off.

Make sure the person you are sending is set to receive AirDrop only from Contacts. In that case, and you are in contact, you must enter the email address or mobile number in the contact to use AirDrop.

If you are not in contact, set the AirDrop receiver to All to receive the file.

You can set the AirDrop Receive setting to Connections Only or Off at any time to control who can see your device and send content in AirDrop.

Using AirDrop

- Open an app, press Share, or Share. When you share a photo from Photos, you can swipe left or right and select multiple photos.

- Touch the AirDrop user * you want to share with. Or you can use AirDrop on your own Apple devices. Do not see the AirDrop user or the other device? Find out what you need to do.

- If the person you are sharing with us in your contacts, a picture with your name will be displayed. If they are not in contact, you will only see their names without a picture.

How to accept AirDrop

When someone shares something with you using AirDrop, a preview alert is displayed. Press Accept or Reject.

If you click Accept, AirDrop will go through the same app you sent it from. For example, photos appear in Photos, and websites open in Safari. Application links open in the App Store, so you can download or purchase the app.

If you give yourself an AirDrop, such as a photo from iPhone to Mac, you will not see the Accept or Reject option - it will be automatically sent to your device. Just make sure both devices are signed in with the same Apple ID

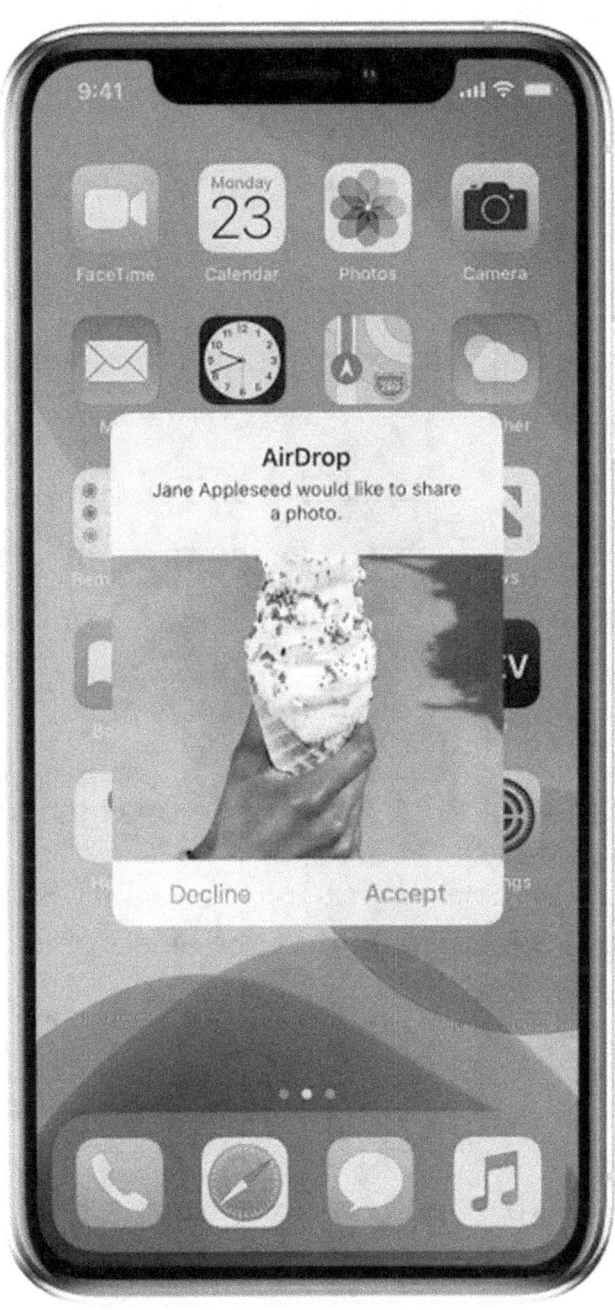

Change AirDrop settings

To choose who can see your device and send content in AirDrop:

- Go to Settings, tap General.
- Press AirDrop, then select an option.
- You can also adjust AirDrop settings in Control Center. This is how:
- On an iPhone X or later, or an iPad or iPad running iOS 12 or later, slide your finger down from the upper right corner of the screen to open the Control Center. On the iPhone 8 or later, slide your finger up from the bottom of the screen.
- Press or hold and hold the Network Settings tab at the top left.
- Press and hold the AirDrop button and select one of the following options:
- Off: You will not receive AirDrop requests.
- Contacts only: Only contacts can see your device.
- Everyone: All nearby Apple devices using AirDrop can see your device.
- If you see the payout option and it does not affect your change:

- Select Settings> Screen time.
- Touch Content and Privacy Restrictions.
- Tap Allowed apps and make sure AirDrop is turned on.

CHAPTER TWELVE

CHANGE NOTIFICATION ON IPHONE 12

In iOS 12, Apple introduced new notification features that offer an expanded set of tools to track and manage alerts faster and more intuitively.

There has been no change in the overall functionality of alerts, but many of these features make it easier to remove alerts, define desired alerts, and make changes on the go.

Grouped alerts
iPhone and iPad owners have been asking for grouped alerts to be returned for years, and in iOS 12, Apple delivered it.

Several alerts from the same app are grouped on the iPhone lock screen, which reduces clutter. Tap a notification set for a specific app to expand them to see all the notifications in the list.

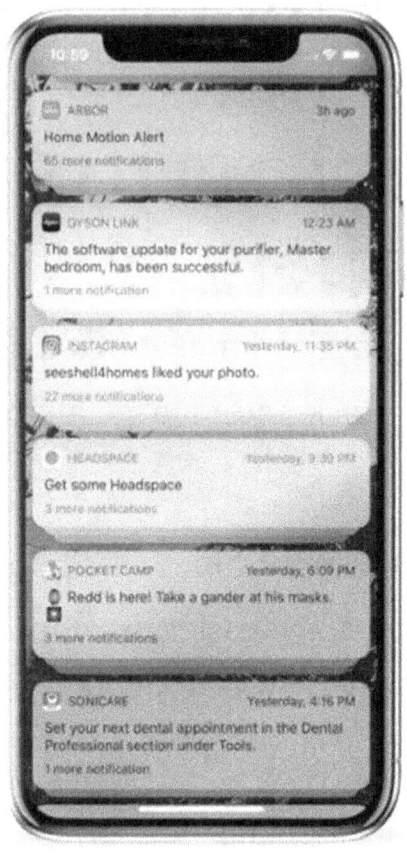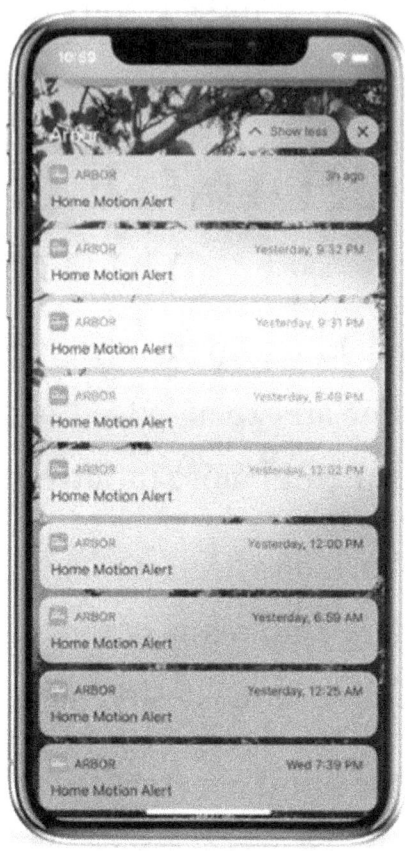

You can delete all notifications at once by pressing "X" next to the notification group or swipe left.

In Settings, you can change the behavior of grouped alerts. Go to Settings> Notifications and tap an app to see the "Notification grouping" settings. Tap this to select "Automatic", "Per-app" or "Off" if you prefer to see all incoming notifications for a specific application, such as Messages.

Auto sorting is mainly done by app, but if you enable this setting, you can get two groups of notifications when email threads go to two different people in Mail, or if you have multiple conversations in Messaging, for example. Or various incoming messages calls.

The application ensures that all alerts from the application are in a pile, without sorting by automatic mode.

Instant setting
Instant Tuning is a feature that allows you to handle annoying alerts directly on the lock screen and provides the ability to turn off app alerts completely or send alerts directly to the alert center.

On any alert on the lock screen or in the notification center when you swipe down, swipe the alert to the left to see the "Manage", "View" and "Delete All" settings.

To see the Instant Tuning options, select "Manage" from the list. Alerts sent to "Live Silent" will be visible in the notification center, but you will not be able to see them on the lock screen, there will be no banner ads and there will be no tags.

To reverse this, tap the notification for the muted app again, follow the same instructions, and select "Shipping Visible." You can also change notification settings in the Settings app, which is also available from the Instant Tuning pop-up window. Disabling, as the name implies, disables alerts for that application completely.

You can also access the Instant Tuning settings by tapping 3D or tapping and holding an alert and selecting the three ellipses. For more information on using Instant Tuning, see detailed description.

On iOS 12, Apple sends you notifications that you will still receive notifications from a particular app if you have received many notifications and have not yet contacted them. In this case, the "Management" section will appear in the alarm, so you can access the instant settings for that program.

Critical alerts
Critical Alerts is a new type of authorization alert in iOS 12 that can override the Do Not Disturb settings to send important required attractions.

These alerts are limited to medical and health information, home security, and public safety. For example, a diabetic may want to set up critical alerts on the blood glucose meter if the blood glucose is low, so the alert will be received even if the Do Not Disturb feature is turned on.

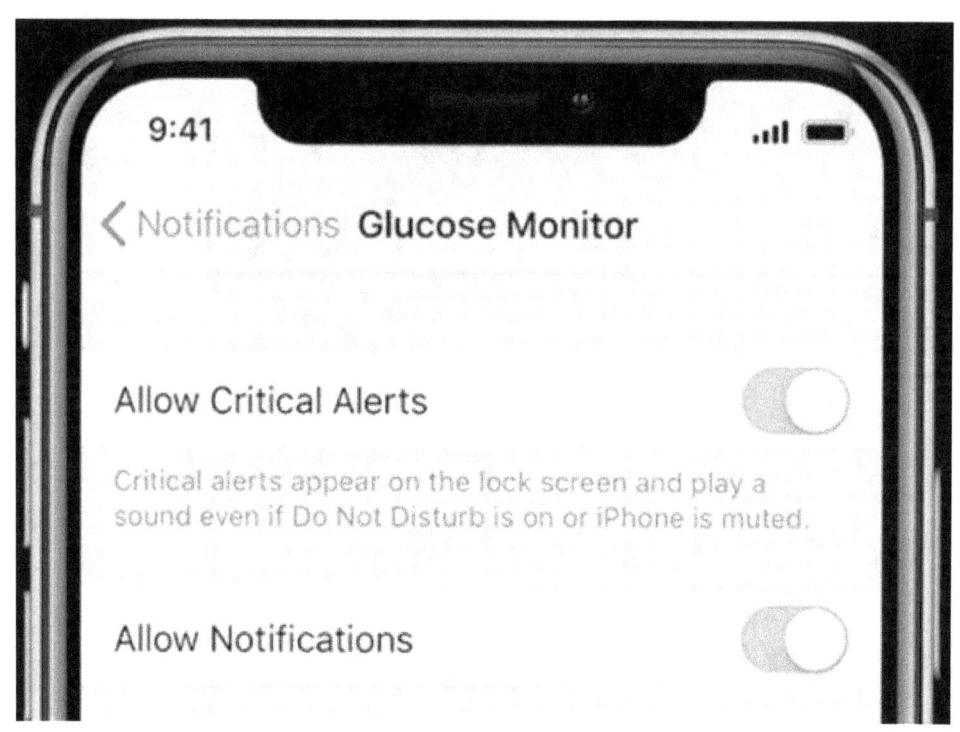

Critical warnings bypass the Do Not Disturb and Ring switches and will always play a sound. They are believed to be confusing and therefore limited to applications that require this type of instant notification.

> **"Glucose Monitor" Would Like to Send You Critical Alerts**
>
> Critical Alerts always play a sound and appear on the lock screen even if your iPhone is muted or Do Not Disturb is on. Manage Critical Alerts in Settings.
>
> Don't Allow   Allow

Developers with applications that are eligible for critical alerts must require an authorization that must be approved by Apple. Users can turn off critical alerts on an application basis, apart from other alerts.

Number of alerts during screen time
Screen Time, Apple's new feature designed to provide devices to monitor the use of iOS devices, keeps track of all the notifications the apps send you and tells you which apps are the noisiest.

This information can help you decide if you want to keep alerts on a particular app or mute an app to reduce interference.

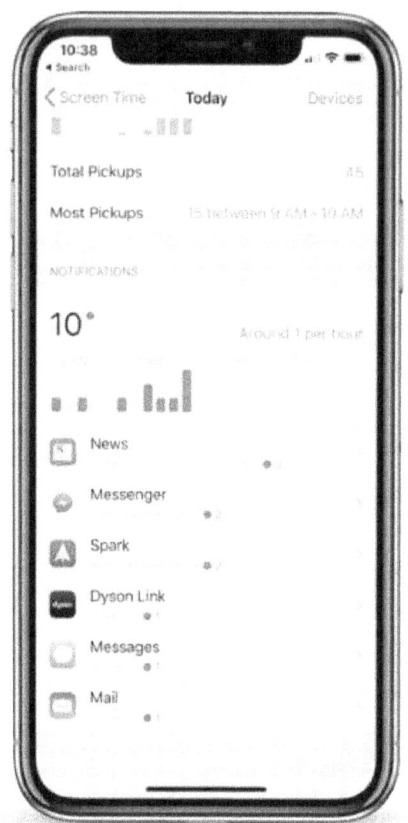 

To access this part of screen time, open the Settings app, select Screen time, select "All devices", and then scroll down. You can see alerts from the last 24 hours or the last seven days.

To learn more about using screen time, check how and how screen time applies to application limits and downtime.

Richer alerts

In iOS 12, app developers can create notifications that can accept user input, so they can interact with notifications in a new way and do more on the lock screen without having to open the iPhone.

For example, with the Instagram app, if the app sends you a notification posted by a friend, you might see the photo and then add an item you want from the notification.

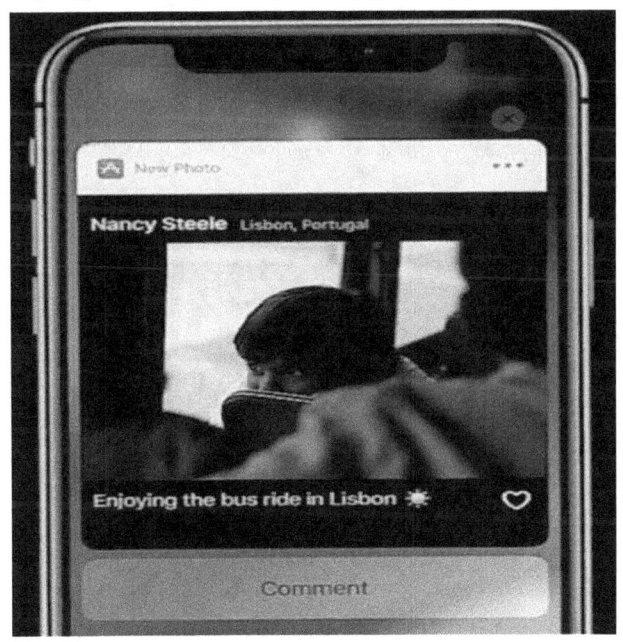

you can touch your heart to like the image, which was not possible in iOS 11.

Rich alerts were available in earlier versions of iOS, but Apple has removed restrictions that previously limited interactive touches.

Turn on Do Not Disturb while driving on iPhone

Do not disturb while driving helps you concentrate on the road. When turned on, text messages and other notifications are muted or restricted. You can ask Siri to read the answers so you do not have to look at the iPhone. Incoming calls are only allowed if the iPhone is connected to CarPlay - a car Bluetooth system - or a handsfree accessory, or if you do not use the Do Not Disturb settings to allow any calls.

WARNING: For important safety information for navigating and avoiding interference in dangerous situations, see Important safety information for iPhone. Do not disturb while driving, it is not a substitute for following all the rules that prohibit disturbing driving.

Switch on Do not disturb while driving

If the iPhone detects that you can drive and you have not selected the Do Not Disturb while driving option, you will be asked if you want to turn it on. Otherwise, you can turn it on manually.

Select Settings> Do Not Disturb.

Scroll down, then tap Activate.

Select when you want to activate the Do Not Disturb feature while driving.

Automatic: When the iPhone detects that it can run.

When connected to car Bluetooth: When iPhone is connected to the car's Bluetooth system.

Manual: When you turn it on in the control center.

Activation using CarPlay: Automatically when iPhone connects to CarPlay.

By adding Do Not Disturb while driving, go to Control Center in Settings> Control Center, and then tap Insert next to Do Not Disturb while driving.

You can receive calls, messages, and alerts when you are a passenger

If Do Not Disturb becomes active while driving when not driving (for example, as a passenger), you can turn it off.

Touch Do Not Disturb while driving on the lock screen.

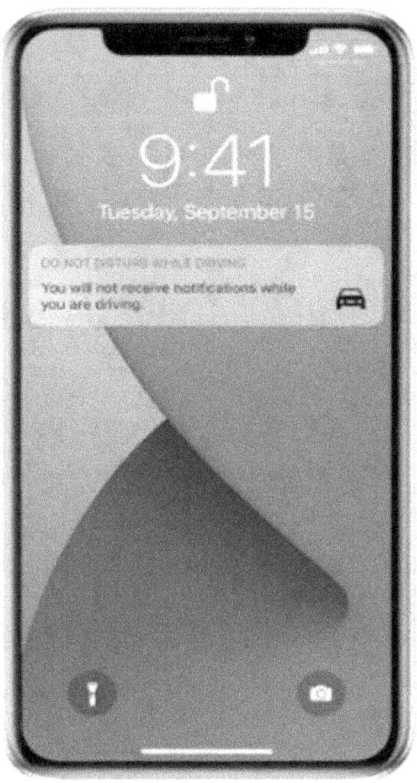

The Do Not Disturb warning while driving on the lock screen.

Press I do not drive.

You can also steal from the bottom of the screen (on an iPhone with Face ID), or press the Home button (for other iPhone models) and then tap Do not drive.

Send an auto-reply message as you drive
If Do Not Disturb while driving is on, it automatically sends a reply to any member of the favorite group. You can change who gets an automatic answer.

Select Settings> Do Not Disturb> Auto Answer.

Choose from the following:

None: Turns off auto answer.

Recent: Sends an automatic reply to everyone you have sent a message to the previous two days, even if it is not in your contacts.

Favorites: Sends an automatic reply to everyone in the favorite group on the phone.

All Contacts: Sends an automatic reply to everyone in Contacts.

When someone responds to the autoresponder message with an "Urgent" message, all the person's additional text arrives at the rest of the station.

Create a custom auto-reply message
Go to Settings and tap Do Not Disturb> Auto Answer

Tap the message to display the keypad, and then enter a new message.

Allow some calls
If your car does not have Bluetooth or CarPlay support, you can allow some calls to be routed.

To allow another call from the same person within 3 minutes: Select Settings> Do Not Disturb and then turn on the call again.

To allow calls from favorites, all contacts, or all: Go to Settings> Do Not Disturb> Allow calls.

Do Not Disturb While Driving Use Location Services to determine if you are driving or near a home, job, or expected destination. Location data collected by Apple for this purpose does not personally identify you. To turn off location services while driving, do not go to Settings> Privacy> Location services> System services, and then turn off location alerts.

Dictated text on iPhone
Switch iPhone on, you can dictate text instead of typing.

Note: Dictation may not be available in all languages, countries, or regions, and features may vary. Mobile data charges may apply. See View or change mobile phone settings on the iPhone.

Turn on dictation
Go to Settings> General> Keyboard.

Turn on Activate dictation.

Dictation of text
Press the Dictate button on the on-screen keyboard and talk.

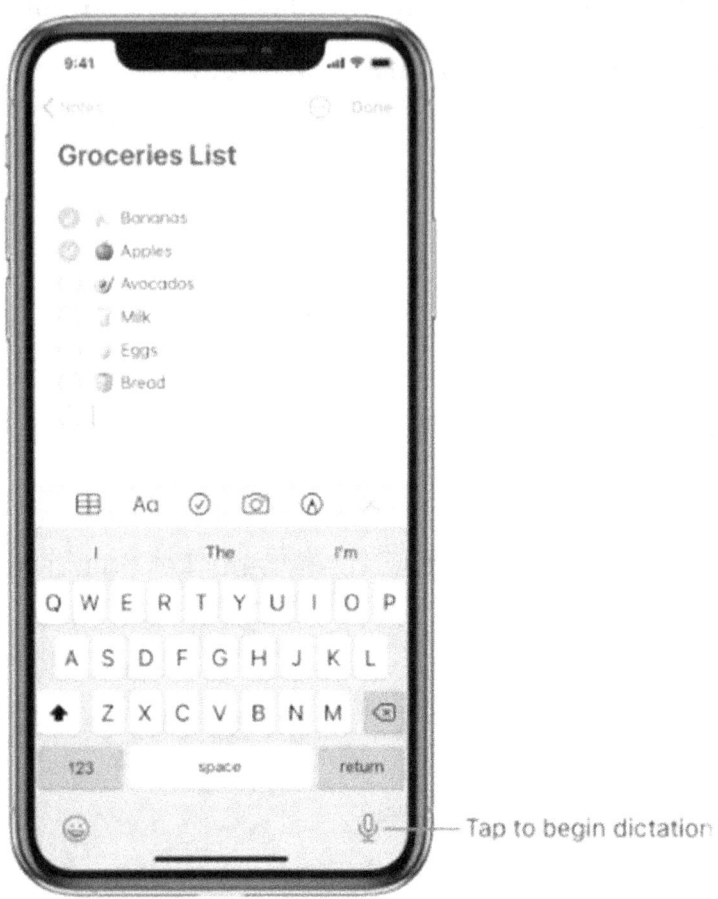

Tap to begin dictation

If you do not see the dictation button, make sure Enable dictation is turned on in Settings> General> Keyboard.

When done, press Keyboard.

The on-screen keyboard that displays the dictator key in the lower right corner. To insert text after dictation, press to place the insertion point, then press Dictation. You can also replace the selected text with dictation.

Format punctuation or text
Say punctuation or formatting while dictating text.

Ex: Dear ann, comma, the check is in the exclamation mark": "Dear ann, the check is in the mail!" Punctuation and formatting commands include:

period

comma

exclamation mark

question marks

dollar sign

open parentheses ... parentheses

quote ... end of quote

new section

new line

colon

semicolon

cap - capitalize on the next word

Caps on ... Caps off - capital letters in every word

all uppercase letters - the next word will be uppercase letters

all caps are exposed ... all caps are exposed - the attached words become large

no caps... no caps off - to make the attached words smal

no word - close the space between the two words (not available in all languages)

no space in ... no words outrun word strings together (not available in all languages)

smiley - for insertion :-)

gloomy - insert :-(

blink - to glue ;-)

Doublecross

CHAPTER THIRTEEN

ADD OR CHANGE KEYBOARDS ON IPHONE

You can turn typing functions such as spell checking on and off; add keyboards for typing in different languages, and change the layout of the monitor or wireless keyboard.

If you add a keyboard to other languages, you can enter two languages without having to switch between keyboards. The keyboard automatically switches between the two most commonly used languages. (Not available in all languages.)

Add or remove a keyboard for another language
Go to Settings> General> Keyboard.

Tap Keyboards, and then do one of the following:

Add a keyboard: Tap Add a new keyboard, then select a keyboard from the list. Repeat to add more keyboards.

To remove a keyboard: Tap Edit, tap Delete next to the keyboard you want to remove, tap Delete, and then tap Done.

To rearrange the keyboard list: Tap Edit, drag the Rearrange button next to the keyboard to a new location in the list, and then tap Done.

If you add a keyboard to another language, the corresponding language is automatically added to the list of preferred language orders. You can view this list and add languages directly in Settings> General> Language & region. By rearranging the list, you can change how applications and websites display text.

Switch to another keyboard
While typing, press and hold Next keyboard, Emoji, or Switch keyboard. Touch the name of the keyboard you want to change.

You can also tap the Next keyboard emoji or Switch the keyboard to switch from one keyboard to another. Keep pressing to access other activated keyboards.

To switch between the magic keyboard, see Switch between the language and the emoji keyboard.

Assign an alternate layout to the keyboard

You can use an alternate keyboard layout that does not match the keys on the keyboard.

Select Settings> General> Keyboard> Keyboard.

Tap a language at the top of the screen, then select an alternate layout from the list.

Use the iPhone to search

iPhone Search is the best place to start all your searches. Search can help you find apps and contacts, search apps like email and messaging, find and open web pages, and quickly launch a web search.

You can choose which apps you want to include in the search results.

- Search offers suggestions and updates the results as you type.
- Select which apps you want to add to Search
- Select Settings> Siri and Search.
- Scroll down, tap an app, and turn View in Viewer on or off.

Search on iPhone

Swipe down from the center of the Home screen.

Tap the search box, then type what you're looking for.

Do one of the following:

To hide the keyboard and display more results on the screen: Tap Go.

Open the suggested app: Tap it.

For more information on a search proposal, tap it and then tap one of the results to open it.

Start a new search: Tap Remove text in the search field. The search screen on the iPhone. At the top is the search box with the search text "apple", and below are the search results for the target text.

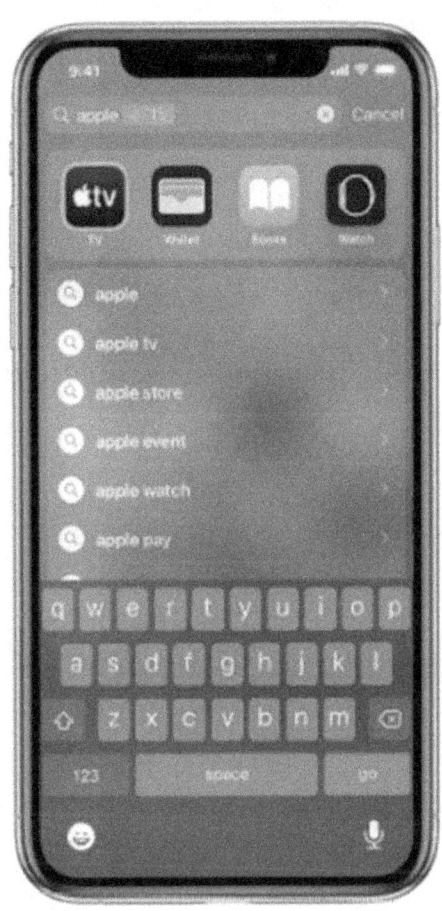

Turn off search suggestions
Select Settings> Siri and Search, and then turn off Search suggestions.

Turn off position services for suggestions

Select Settings> Privacy> Location services.

Tap System Services, and then turn off location-based suggestions.

Search applications

Many apps have a search box or search button so you can find something in the app. For example, you can search for a specific place in Maps.

Press the search field or search button in an application.

If you do not find a search box or button, slide your finger down from the top.

Enter the search, then tap Search.

Add a dictionary

You can add dictionaries on the iPhone that can be used in searches.

Select Settings> General> Dictionary.

Select a dictionary.

CHAPTER FOURTEEN

CUSTOMIZE THE CONTROL CENTER ON IPHONE

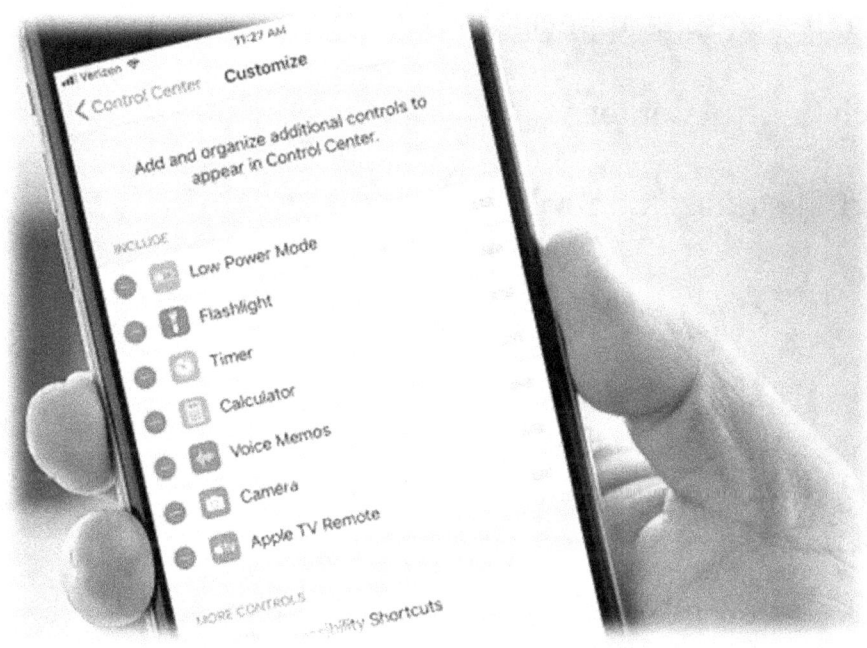

Just as you arrange gadgets in today's view, you can select specific iPhone features so that they are quickly accessible from the control center. It also lets you remove Control Center features that you never use and replace them with something better. How to customize the control center on the iPhone

Customize the control center

First, slide down from the top right corner of the screen (or up from the bottom of the screen if you have an iPhone 8 or earlier) to open the Control Center.

Open the Settings application.

Touch Control Center.

Select Customize Controls.

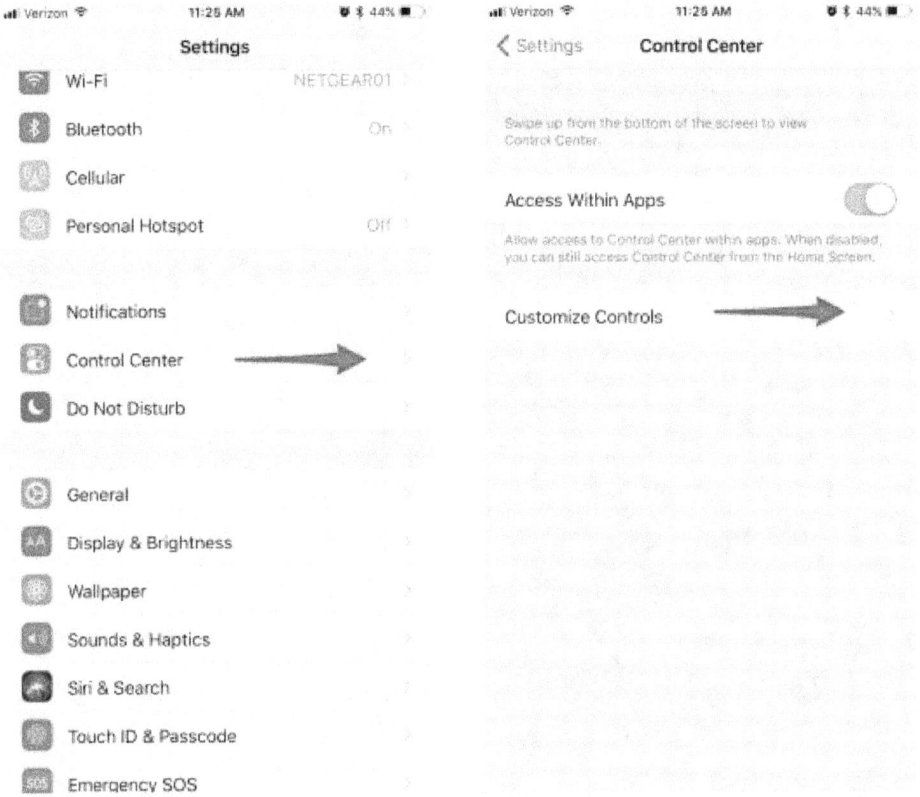

To remove a controller, tap the red circle next to the function you want to remove.

Touch the green circle under Additional Controls to add the function to the Control Center.

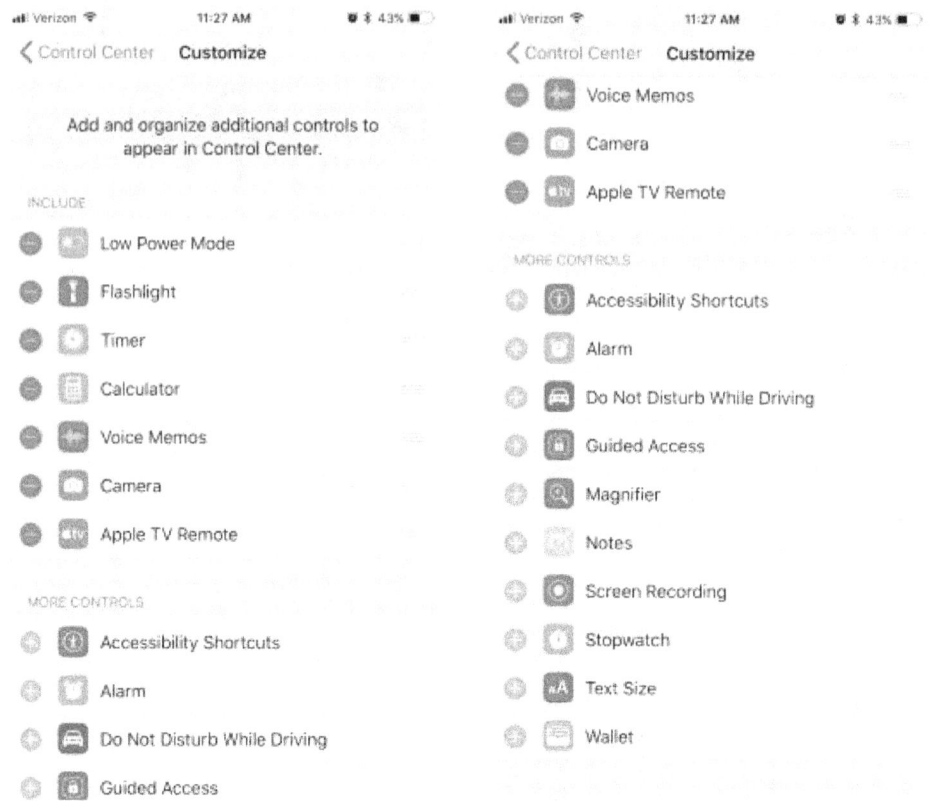

- Under Include, you can tap and hold the three horizontal lines to the right of each function to rearrange the functions and their appearance in the Control Center.

I like the Apple TV remote control option, but what I add and use best is, of course, low power mode.

Add widgets to the iPhone Home screen

Today, gadgets instantly show you data from your favorite apps - today's headlines, weather, calendar events, and more. You can add gadgets to the home screen to keep this information convenient.

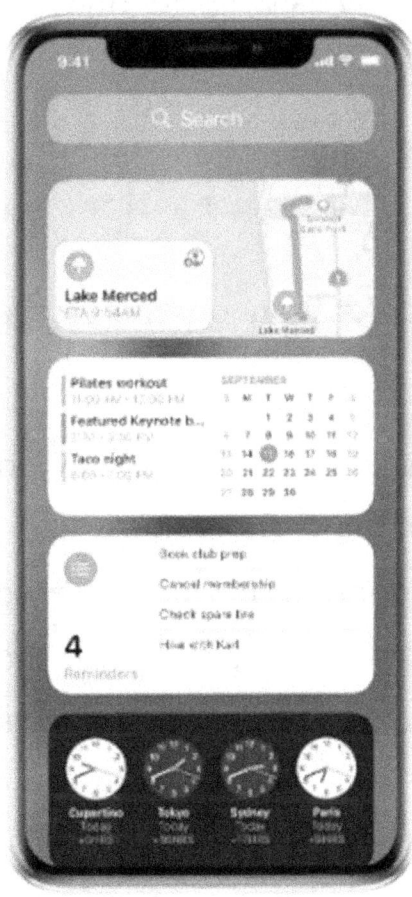

View widgets on iPhone today, including Maps, Calendar, Reminders, and Clock.

Open today's view
Swipe right from the left edge of the home screen or lock screen.

Move a device from the day view to the home screen
Open the Today view and scroll or search to find the module you want.

Press and hold the module until you start laughing, then drag it from the right side of the screen.

Drag the device to the desired location on the Home screen, then tap Done.

Tip: A widget called a smart stack (one with dots next to it) is a widget that uses information such as time, place, and activity to display the most appropriate gadget at the right time of day. You can add a smart package to the home screen and then scroll through it to see the widgets in it.

Add a gadget to the Home screen
From the Home screen, where you want to add the device, then press and hold the wallpaper on the Home screen until the apps start dragging.

Tap the Add Widget button at the top of the screen to open the widget gallery.

Scroll or search for the desired module, tap it, and then drag the size settings to the left.

Different sizes show different information.

When you want the size, tap Add widget and then Done.

Tip: A widget called a smart stack (one with dots next to it) is a widget that uses information such as time, place, and activity to display the most appropriate gadget at the right time of day. You can add a smart package to the Home screen, and then scroll to see the widgets in it.

Edit module
You can customize most modules to display the information you want. For example, you can edit a weather gadget to see a forecast for your location or another area. Or you can customize your smart pile to

move through the modules based on things like activity, time of day, and more.

Tap and hold a gadget on the Home screen to open the Quick Actions menu.

Tap Edit widget if it appears (or Edit stack if it is a smart stack), and then select Settings.

For example, for a weather widget, you can tap Location and then select a forecast location.

For Smart Stack, you can turn Smart Rotate on and off and rearrange the widget by dragging the Rearrange button next to it.

Touch the home screen.

Remove a device from the Home screen
Press and hold the module to open the quick action menu.

Touch Remove Widget (or Remove Stack), then touch Remove.

Give access to the day view when iPhone is locked
Go to Settings> Face ID and Password (for iPhone with Face ID) or Touch ID and Password (for other iPhone models).

Enter your password.

Turn on the day view (under Activate locked).

Select iPhone settings to travel

When traveling with an iPhone, select settings that minimize mobile dosing and meet airline requirements. See View or change mobile phone settings on the iPhone. Some airlines allow you to keep your iPhone turned on when you switch to airplane mode. In Airplane mode, you can not make or use Bluetooth, but you can listen to music, play games, watch videos, or use other applications that do not require networking or dial-up connection.

Turn on airplane mode
Open the Control Center and press the Flip Mode rocker.

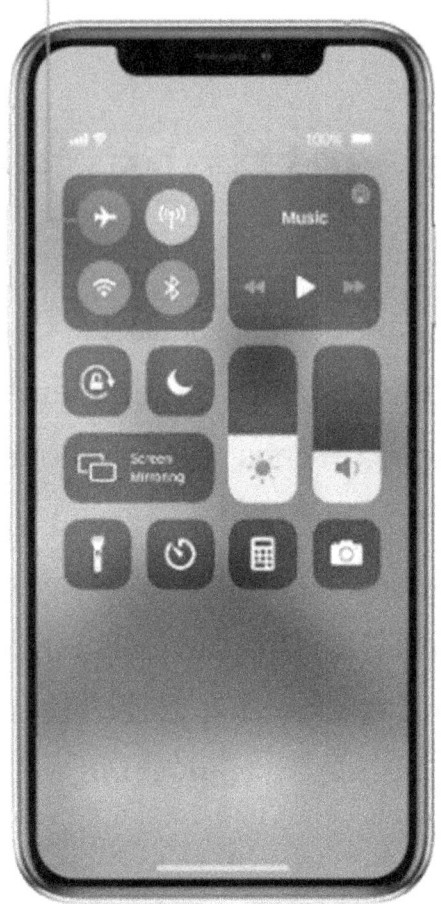

Tap to turn on airplane mode.

You can also turn flight mode on or off in Settings. When airplane mode is on, the airplane mode icon appears in the status bar.

In airplane mode, turn on Wi-Fi or Bluetooth
If your airline allows it, you can use Wi-Fi or Bluetooth in airplane mode.

Open the Control Center and then turn on airplane mode.

Press the Wi-Fi power button (for Wi-Fi) or the Bluetooth power button (for Bluetooth).

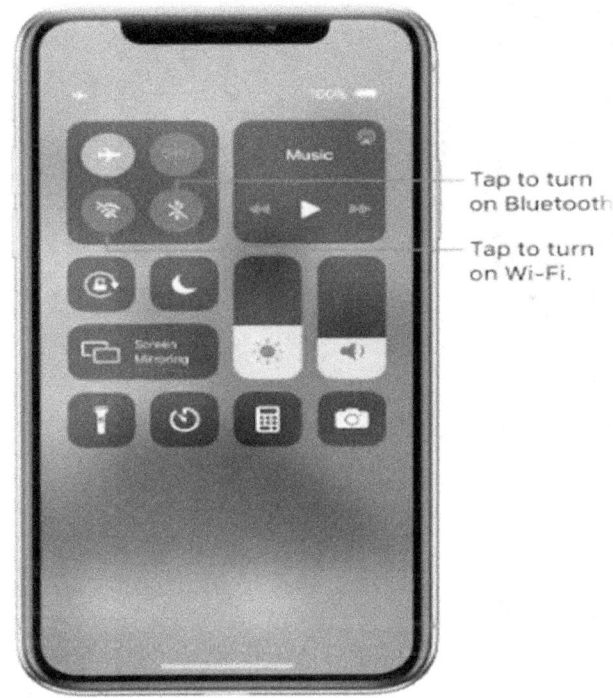

If you turn on Wi-Fi or Bluetooth in airplane mode, you will return the next time you return to airplane mode. You can turn it off again in the Control Center.

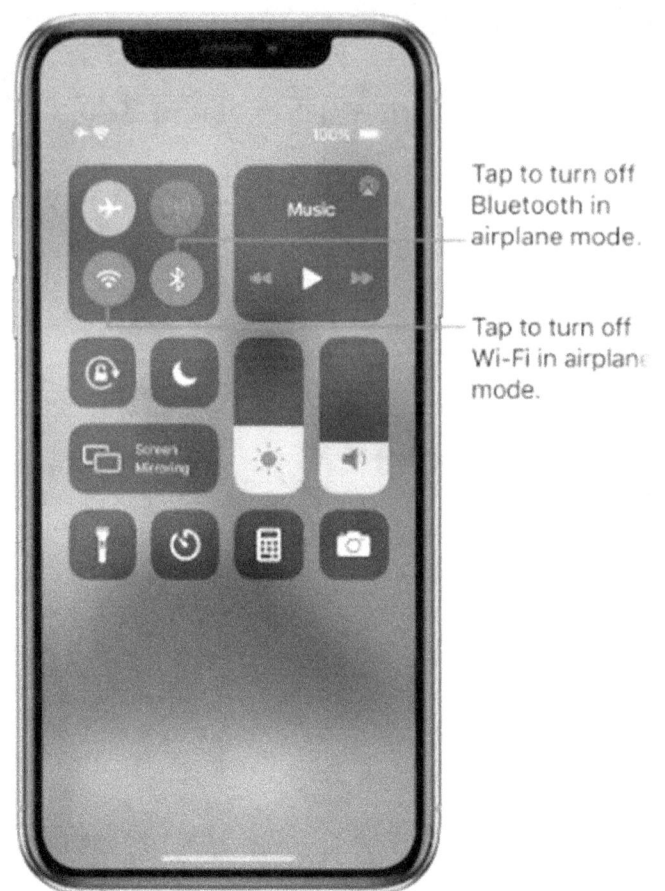

Use screen time on iPhone, iPad, or iPod touch

Using screen time, you can access real-time reports on how much time you spend on your iPhone, iPad, or iPod touch, and set limits on the data you want to manage.

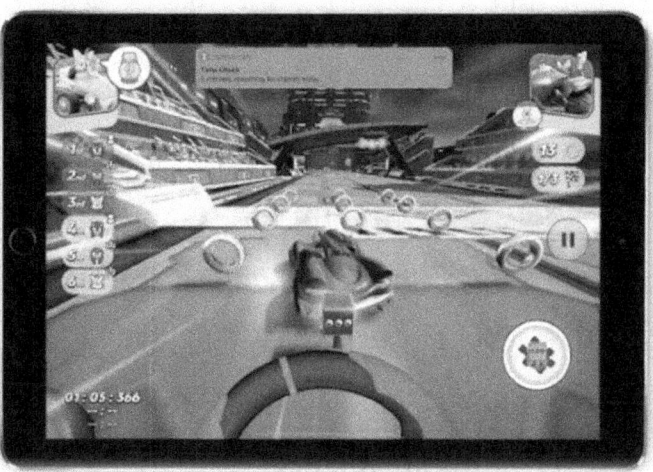

Screen time lets you see how much time you and your kids spend on apps, websites, and more. This way, you can make more informed decisions about how to use your devices and set boundaries if you want. Read on to learn how to turn on screen time, view the report, and set boundaries, and manage your child's device.

Turn on screen time
Select Settings> Screen time.

Tap Turn on the screen.

Press Continue.

Select This is My [device] or This is My Child's [device].

You can get a report on the use of your device, apps, and websites at any time.

If this is your child's device, you can set screen time and create settings directly on your device, or use Family Sharing to configure your child's device from your device. Once you have set up your child's device, you can use Family Sharing to view reports and change settings directly from your device.

You can also use screen time to create your password to secure your settings so that only you can extend the time or make changes. Be sure to choose a password that is different from the code used to unlock the computer. To change or turn off the password on your child's device, go to Settings> Screen time and press [child name]. Then tap Change screen time password or Turn off-screen time password and confirm the change with Face ID, Touch ID, or device password.

If you forget your screen time password, update your device to the latest iOS or iPadOS, and then reset your password. If you cannot update your device, delete and enter a new one to clear the password and select a new one. Restoring a device from a backup does not remove the password.

Enter a screen time password

Enter a password so that only you can change the screen time settings and allow more time when the application limits expire. If you are a parent, use this feature to set content and privacy restrictions for your child.

If you use Family Sharing to manage your child's account, follow these steps:

Touch Settings> Screen Time.

Scroll down and select the child's name under Family.

Tap Turn on the screen, then tap Continue.

Enter downtime, application limits, content, and privacy with all the restrictions your child wants, or tap Not Now.

Touch Use Screen Time Code, and then enter the code when prompted. Re-enter the password to confirm.

Enter your Apple ID and password. This will reset the screen time code if you forget it.

If you are not using Family Sharing to manage your child's account, follow these steps:

Make sure it is on the device that the child is using.

Touch Settings> Screen Time.

Tap Turn on the screen, then tap Continue.

Select This is the child's [device].

Enter downtime, application limits, and content and privacy with all the restrictions your child wants, or tap Not Now.

Touch Use Screen Time Code, and then enter the code when prompted. Re-enter the password to confirm.

Enter your Apple ID and password. This will reset the screen time code if you forget it.

See your report and set boundaries
- Screen time provides a detailed report on device usage, open apps, and websites you've visited when you want to see it. Just go to Settings> Screen time and tap Show all activities below the graph. From here, you can see your usage, set limits on your

most-used apps, and see how many times a device has detected or received alerts.

- If you have turned on device sharing, you can see the general usage of devices logged in with Apple ID and password.

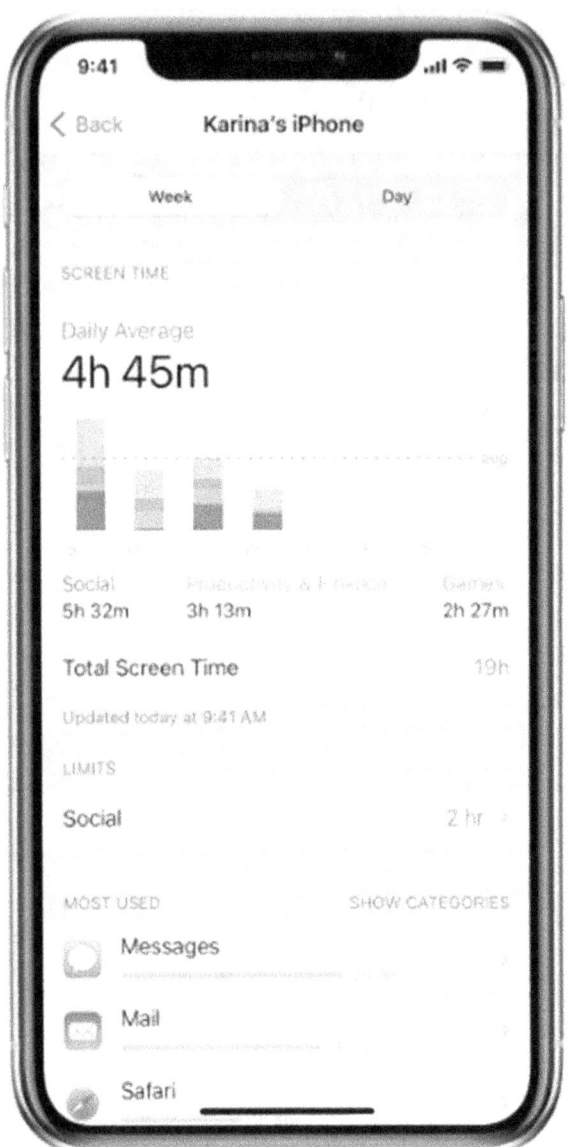

You can manage the following settings:

Downtime
Think of it as a snooze of time on screen. If you schedule downtime in Settings, only the phone calls and apps you

have activated are available. Downtime applies to all devices with screen time, and you will be reminded five minutes before departure.

App limits

You can set daily limits for application categories using application limits. For example, you may want to see productivity apps at work, but not social networks or games. App restrictions are updated every day at midnight and you can delete them at any time.

Communication boundaries

Make sure your kids can communicate with you - during the day and downtime. These restrictions apply to phone, FaceTime, Messages, and iCloud contacts. Here you can also decide and manage which available connections are available on the Apple Watch paired via Family Setup. Communication with known emergency numbers identified by your mobile operator on iPhone or Apple Watch is always allowed. You need to enable iCloud contacts to use this feature.

always allowed

You may want to access certain applications, even if it is shutdown or if you have set the application limit for all applications and categories. Phone, Messaging, FaceTime, and Maps are always on by default, but you can delete them if you want.

Content and privacy restrictions
You decide what type of content is displayed on your device. Block inappropriate content, purchases, and downloads, and adjust your privacy settings using content and privacy restrictions.

Spend screen time with your family
With family sharing, you can share music, movies, apps, and more with your family - and now screen time works. You can view reports and family child settings at any time, directly on your device.

If you already belong to a family group, go to Settings> Screen time and tap the child's name. If you want to create an Apple ID for your child, go to Settings> [your name]> Family Sharing> Screen Time.

Or if you do not already know family sharing, tap Set family screen time and follow the instructions to add your child and create a family. You can add family members at any time from the family sharing settings.

To use screen time with family sharing, you must be a family organizer or parent/guardian in the family group, on iOS 12 and later, or iPad. Your child must be under 18, with their own Apple ID in the family group, and on iOS 12 and later or iPad.

CHAPTER FIFTEEN

SET CONTENT AND PRIVACY RESTRICTIONS

Go to Settings and tap Screen time.

Press Continue and select "This is my [device]" or "This is my child [device]".

If you are the parent or guardian of your device and want to prevent another family member from changing your settings, tap the Use screen time tag to create a password, then enter it again to confirm. In iOS 13.4 or later, after confirming the password, you will be asked to enter your Apple ID and password. This will reset the screen time code if you forget it.

If you set the screen time on your child's device, follow the instructions until you get the parent code and the code. Re-enter the password to confirm. In iOS 13.4 or later, after confirming the password, you will be asked to enter your Apple ID and password. This will reset the screen time code if you forget it.

Touch Content and Privacy Restrictions. If prompted, enter your password, then turn on content and privacy.

Be sure to choose a password that is different from the code used to unlock the computer. To change or turn off the password on the child's device, touch Settings> Screen time> [child name]. Then tap Change screen time password or Turn off-screen time password and confirm the change with Face ID, Touch ID, or device password.

Prevent purchases from iTunes and the App Store
You can also prevent your child from installing or uninstalling apps, making in-app purchases, and more. To prevent purchases or downloads from iTunes and the App Store:

- Go to Settings and tap Screen time.
- Touch Content and Privacy Restrictions. If prompted, enter your password.
- Tap iTunes and App Store purchases.
- Select an option and set it to Disable.
- You can also change the password settings for further purchases from iTunes and the App Store or Book Store. Follow steps 1-3. Step, and then select Always Request or Do Not Require.

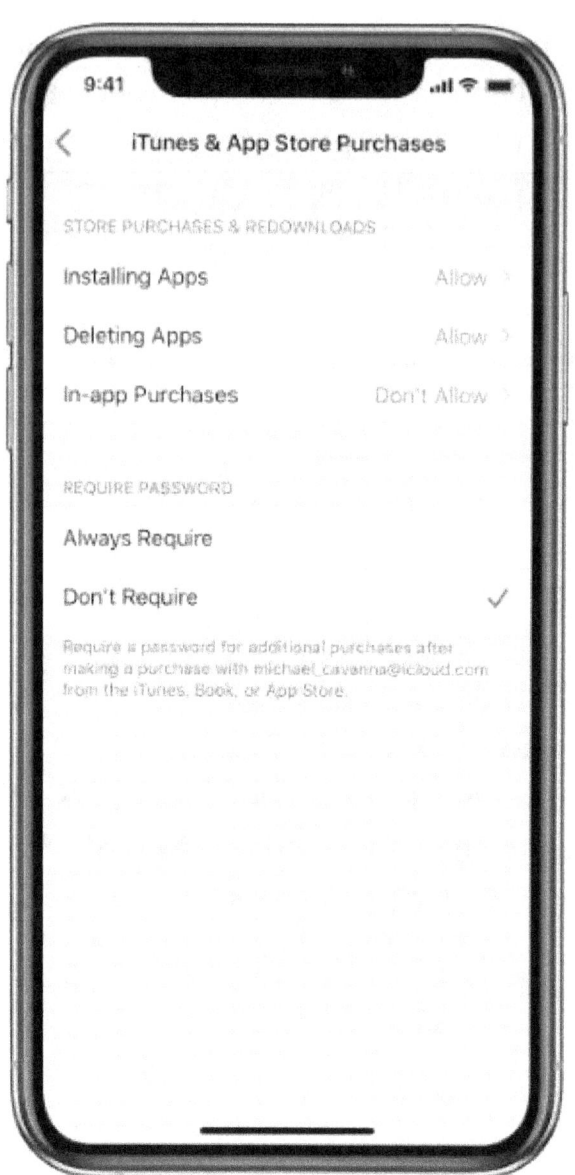

Enable built-in applications and features

You can restrict the use of embedded applications or services. If you turn off an app or feature, it will not be deleted, it will only temporarily hide it from the Home screen. For example, if you turn off Mail, Mail only appears on the Home screen when you turn it on again.

- Change permitted applications:
- Select Settings> Screen time.
- Touch Content and Privacy Restrictions.
- Enter the screen time password.
- Tap Allowed Apps.
- Select the programs you want to activate.

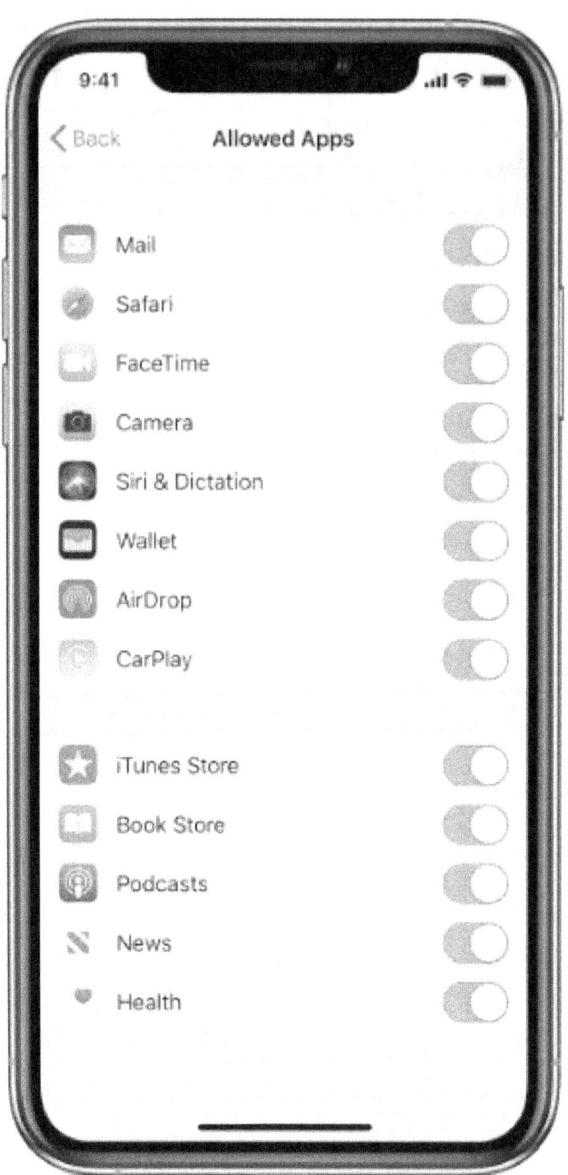

Prevent explicit content and content rating

You can also prevent music with explicit content from being played, as well as movies or TV series with a specific rating. Applications also have configurable rankings with content restrictions.

Restrictions on Express Content and Content Classification:

- Go to Settings and tap Screen time.
- Touch Content & Privacy Restrictions, then touch Content Restrictions.
- Select the desired settings for each function or setting under Allowed Store Content.
- Here are the types of content you can limit:
- Ranking: Select the country or region in the classification section to automatically apply appropriate content ratings for that region
- Music, podcasts, and news: Prevent music, music videos, podcasts, and explicit content from being played
- Music Videos: Prevents you from searching for and watching music videos
- Music profiles: Prevent you from sharing what you're listening to with your friends and not seeing what you're listening to
- Movies: Prevent movies with certain ratings
- TV shows: You can block TV shows with a specific rating
- Books: You can block certain categories of content
- Programs: You can block applications in a given rating

Prevent web content
IOS can automatically filter the content on your site to restrict access to adult content between Safari and the

apps on your device. You can add specific sites to an approved or blocked list, or restrict access to only approved sites. Move these files:

Select Settings> Screen time.

Touch Content and Privacy Restrictions and enter the screen time password.

Tap Content Restrictions, then tap Web Content.

Select Unlimited Access, Restrict Adult Sections, or Allowed Sites Only.

Depending on the access you provide, you may need to add information, such as the site you want to restrict.

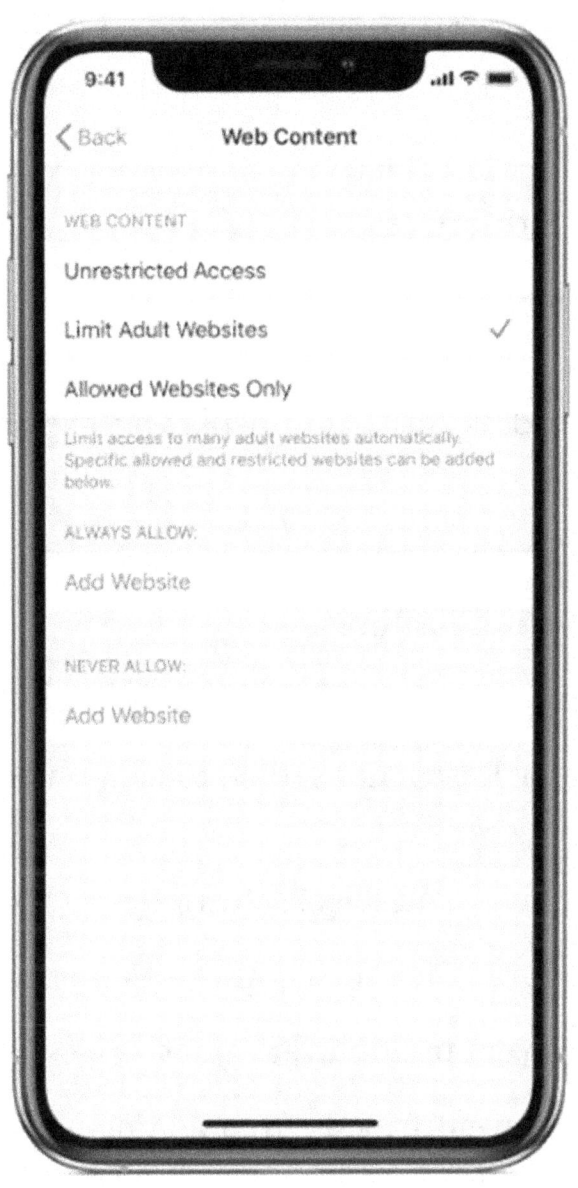

Restrict Siri Internet Search

Limitations on Siri Features:

Go to Settings and tap Screen time.

Touch Content & Privacy Restrictions, then touch Content Restrictions.

Scroll down to Siri and select settings.

You can restrict these Siri services:

Internet content: you can prevent Siri from asking questions online

Explicit language: prevents Siri from displaying explicit language

Limit the game center

Limitations of Game Center Features:

Go to Settings and tap Screen time.

Touch Content & Privacy Restrictions, then touch Content Restrictions.

Scroll down to Game Center and select settings.

You can restrict the following features in Game Center:

Multiplayer Games: Prevent multiplayer games

Add friends: You can prevent friends from adding you to the Game Center

Screen recording: Prevents the screen and sound from being captured

Allow changing the privacy settings

Device privacy settings allow applications to access information stored on the device or hardware features. For example, you can let a social networking application request access to use the camera, which allows you to take and upload photos.

- To allow changes to your privacy settings:
- Go to Settings and tap Screen time.
- Touch Content and Privacy Restrictions. If prompted, enter your password.
- Tap Privacy, then select the settings you want to restrict.
- Here you can limit the following:
- Location services: Close the settings to allow apps and websites to use your location
- Contacts: Prevent applications from accessing Contacts
- Calendars: Prevents applications from accessing Calendar
- Reminders: Prevents apps from accessing Reminders
- Photos: Prevents apps from requesting access to your photos
- Share your location: Close your location sharing settings in Messaging and find your friends
- Bluetooth sharing: Prevents devices and applications from sharing data via Bluetooth

- Microphone: Prevents applications from requesting access to the microphone
- Speech recognition: Prevents applications from accessing speech recognition or dictation
- Advertising: Prevents you from changing your ad settings
- Media and Apple Music: Prevents apps from accessing your photos, videos, or music library

Allow changes to other settings and services

You can allow changes to other settings and services, as well as changes to privacy settings.

Go to Settings and tap Screen time.

- Touch Content and Privacy Restrictions. If prompted, enter your password.
- In the Allow changes section, select the services or settings for which you want to allow changes, and then click Enable or Disable.
- Here are some features and settings for activating changes:
- Password changes: prevents password changes
- Account changes: Prevents you from changing account and password settings
- Changes to mobile data: prevents you from changing mobile data settings
- Volume limit: Prevents you from changing the volume settings for safe listening

- Do not disturb while driving: Do not disturb while driving
- TV provider: Prevent the TV provider's settings from changing
- Background app activities: Prevents apps from running in the background

CHAPTER SIXTEEN

ASK SIRI ON IPHONE

Talking to Siri is a quick way to get things done. Ask Siri to translate a phrase, set a timer, search for a place, report weather, and more. The more you use Siri, the better you know what you need.

To use Siri, your iPhone must be connected to the Internet. Mobile charges may apply.

Adjust Siri

If you did not set up Siri the first time you set up iPhone, go to Settings> Siri and Search and do one of the following:

To summon Siri with your voice: Turn on "Hello Siri".

To call Siri with one button: Tap Press Page to Siri (iPhone with Face ID) or Press Home to Siri (iPhone with Home button).

To change more Siri settings, see Change Siri settings on iPhone.

Invite Siri with your voice

When you call Siri with your voice, Siri answers loudly.

Say "Hi Siri" and ask a question or do an assignment for Siri.

For example, say something like "Hi Siri, how is the weather today?" or "Hi Siri, set the alarm to 8 in the morning"

To ask a question to Siri or perform another task, say the word "Hi Siri" again or press Listen.

To prevent the iPhone from responding to a "Hello Siri" message, place the iPhone face down, or go to Settings> Siri and Search and turn off "Hello Siri".

You can also say "Hello Siri" to call Siri while using AirPods Pro or AirPods (2nd generation). See Using Siri with AirPods on iPhone.

Call Siri with a button

When you call Siri with a button, Siri answers loudly when iPhone is in call mode and silently when iPhone is in silent mode. See Silent iPhone in Ring Mode or Silent Mode. To change this, see Change Siri's response.

Do one of the following:

On an iPhone with face ID: Press and hold the side button. On iPhones with a Home button: Press and hold the Home button.

EarPods: Press and hold the center or call key.

CarPlay: Press and hold the voice command button on the steering wheel, or press and hold the Home button on the CarPlay Home screen. (See Using Siri to Control CarPlay.)

Siri Eyes Free: Press and hold the voice command button on the steering wheel.

When Siri appears, ask a question or do a task for you.

For example, say something like "What is 225 percent?" or "Set the timer to 3 minutes."

To ask a question to Siri or perform another task, tap Listen.

You can invite Siri to AirPods by holding down or double-clicking. See Using Siri with AirPods on iPhone.

Correct if Siri is misunderstood
Write about your request: Tap Listen and say your request differently.

To clarify part of the request: Press Listen and repeat the request with words that Siri did not understand. For example, say the word "Call" and then enter the person's name.

Change a message before sending it: Say "Change".

Edit the request with text: If you see your request on the screen, you can edit it. Tap the application, then use the on-screen keyboard.

Write instead of talking to Siri
Go to Settings> Availability> Siri and turn on Type for Siri.

To make a request, call Siri and use the keyboard and text box to ask Siri a question or perform a task for you.

Find out what Siri can do on the iPhone

You can use Siri on the iPhone to get more information and perform tasks. Siri and the answer are displayed at the top of the ongoing activities, so you can refer to the information displayed on the screen.

- Siri is interactive. When Siri displays a web connection, you can tap it to see more information in the default browser. When the answer on the Siri screen includes buttons or controls, you can tap them to do more. And you can press Siri again to ask another question or perform an additional task.
- Find answers to your questions: Find information online, get sports scores, arithmetic calculations, and more. Say something like "Hi Siri, what's causing the rainbow", "Hi Siri, what was yesterday's Orioles game result," or "Hi Siri, what's the derivative of cosine x?"
- Perform tasks with iPhone applications: Use Siri to control voice applications. For example, to create an event in the calendar, say something like "Hi Siri, make an appointment with Gordon at 9 am", or if you want to add an item to Reminders, say something like "Hi Siri, add courgette and garlic on my menu.

- Send and reply to messages: Say something like "Hi Siri, send a message to Eliza about what's going on tomorrow" or "Hi Siri, reply, that's good news." You can even use Siri to send voice messages.
- If you connect AirPods (AirPods Pro and AirPods Generation 2) to iPhone and a message arrives, Siri will read the message for you, even if the iPhone is locked. Siri listens after reading the messages so you can reply without saying "Hi Siri". See Using Siri with AirPods on iPhone.

- Translate language: Say something like "Hi Siri, how do you say thank you in Mandarin?" or "Hi Siri, what languages can you translate?

- Play a radio station: Say something like "Hi Siri, play Wild 94.9" or "Hi Siri, tune in to ESPN radio"
- Let me show you more examples of Siri: Say something like: "Hi Siri, what can you do?
- More examples are shown in this guide. To learn more about Siri, visit Siri.

CHAPTER SEVENTEEN

SHARE YOUR VOICE WITH IPHONE AIRPODS AND BEATS

While using AirPods or compatible Beats earphones or headphones, you can share your listeners with a friend who also uses an AirPod or compatible Beats product. sets of audio devices must be paired with the iPhone, iPad, or iPod touch. (Supported models; requires iOS 13.1, iPad 13.1 or later.)

- Start sharing audio when charging your friend's AirPods or Beats headsets
- You need to connect an AirPods or Beats product to your iPhone and have your friend's AirPods or Beats earphones in your case.
- While using AirPods or a Beats product, place the iPhone in your friend's open charge.
- On an iPhone, tap Share temporary audio.

- Follow the on-screen instructions and your friend can remove the AirPods or Beats headset from the case.

Share your voice with your friend Beats headphones
You need to connect AirPods or Beats to iPhone and turn on your friend's Beats headphones.

Ask your friend to briefly press the power button on the headset (in less than 1 second).

While using AirPods or a Beats product, place the iPhone on your friend's headphones.

On an iPhone, tap Share audio temporarily, and then follow the onscreen instructions

Share the sound if your friend is wearing AirPods or Beats

If your friend is wearing AirPods or a Beats product connected to an iPhone, iPad, or iPod touch, you can share the audio played on the iPhone.

Use AirPods or a Beats product.
Press the Play Target button on the iPhone in the Now Playing controls, either in the app you are listening to or on the lock screen.

Or open the Control Center, press and hold the playback controls in the upper right corner, then press Play Target.

Touch Share audio (under the name of the headset or headphones).

Take the iPhone closer to your friend's iPhone, iPad, or iPod touch.

Tap Share audio on the iPhone.

Ask your friend to tap Connect on your device.

Adjust the volume of each toolkit
If you and a friend share audio from iPhone using AirPods or Beats, you can set different volume levels for each device.

Open the Control Center on your iPhone and hold down the volume control.

Drag the individual volume sliders.

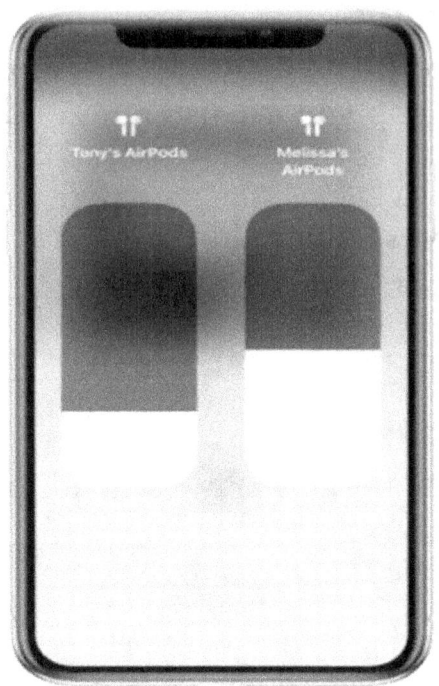

Stop sharing audio

On iPhone, tap the Play Destination button on the Now Playing, Lock Screen, or Control Center screen, and then tap your friend's AirPods or Beats product name to turn off the connection.

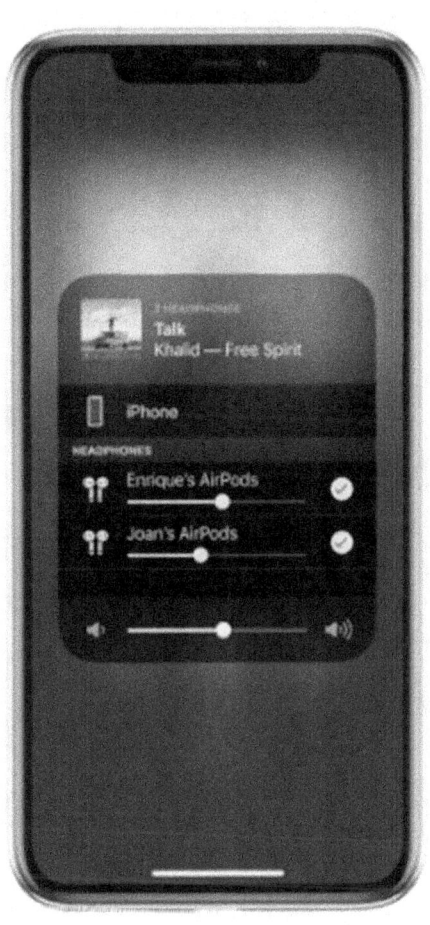

CHAPTER EIGHTEEN

COLLECT HEALTH AND FITNESS DATA IN THE IPHONE APP FOR IPHONE

With the Health app, you can keep track of your daily steps and stairwells. You can manually add other data, such as weight and caffeine intake, and track additional data with other applications (such as nutrition and exercise apps) and health-compatible devices (such as Apple Watch, AirPods, weight, and blood pressure monitors).).

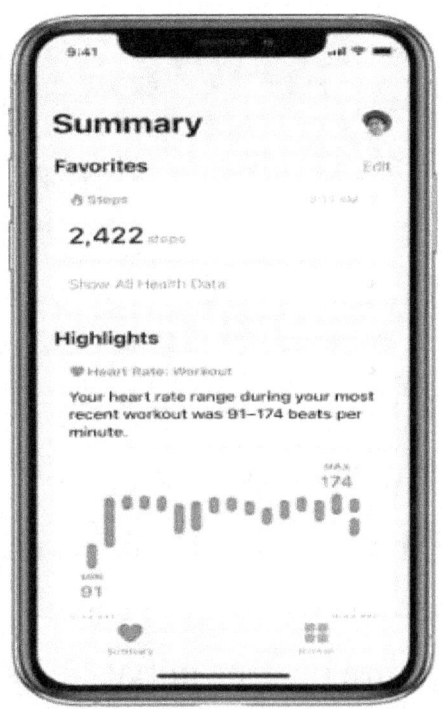

The summary screen shows the steps as a favorite category. Under highlights, you can see information about your last workout on the screen.

Manually update your health profile
When you first open healthcare, you should ask for a health profile that contains basic information such as date of birth and gender. If you do not provide all the required information, you can update your profile later.

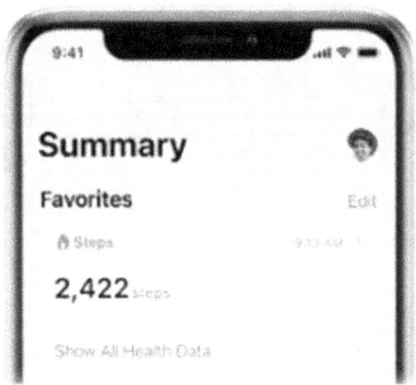

Tap the profile picture or your initials at the top right.

The summary screen shows the profile picture in the upper right corner.

If you do not see the profile picture or your initials, press the Summary or Browse button at the bottom of the screen and scroll to the top of the screen.

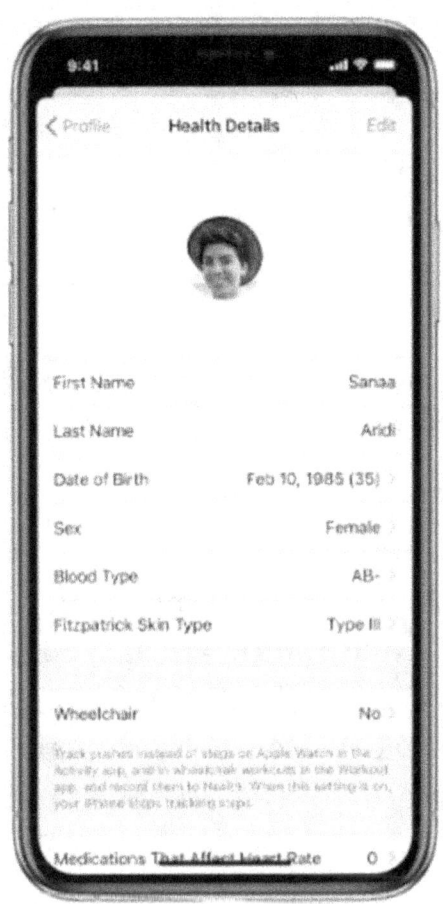

Touch Health Details, then touch Edit.
- The Health Information screen for a 35-year-old woman.
- Tap a field, make a change, and then tap Done.
- Manually add data to a health category
- Tap the Browse button in the lower right corner to display the health category screen, then do one of the following:
- Tap a category. (Scroll down to see all categories.)

- Tap the search box, then enter a category name (such as body size) or a specific type of data (such as weight).
- If you do not see the health category screen, press the Browse button in the lower right corner again.
- Press the Details button for the data you want to update.
- Tap Add data at the top right of the screen.
- Enter your details, then tap Add or Done at the top right of the screen.

Collect data from other sources

From Apple Watch: After pairing the iPhone with the Apple Watch, the Apple Watch automatically sends periodic heart rate measurements to Healthcare. You can also choose to also set up Apple Watch to send activity metrics, noise levels, and more to Health. See Apple Watch User Guide.

- From headphones: Once you have connected EarPods, AirPods, and other compatible headphones to the iPhone, the sound level of the headphones are automatically sent to Health.
- From an app downloaded from the App Store: While setting up the app, you can allow yourself to share data with Health.
- From another device: Follow the device installation instructions.
- If it is a Bluetooth device, you need to pair it with the iPhone. Follow the instructions that came with

the device to put it in discovery mode, select Settings> Bluetooth and turn it on.
-

Set reminders on iPhone

In Reminders, you can easily create and organize reminders to keep track of everything in life. Use it for shopping lists, work projects, homework, and everything else you want an overview of. Create subtasks, set flags, add attachments, and choose when and where you want to receive reminders. You can use smart lists to organize your reminders automatically.

A reminder list appears on the Reminders screen. The new reminder button is located in the lower-left corner.

Keep reminders up to date on all your devices using iCloud

Go to Settings> [your name]> iCloud and turn on Reminders.

iCloud reminders - and any changes you make to them - appear on your iPhone, iPad, iPod touch, Apple Watch, and Mac, where you're signed in with your Apple ID.

Note: If you've used Reminders on iOS 12 or earlier, you may need to update your iCloud reminders to use features like attachments, flags, subtasks, grouped lists, list colors, and icons. To refresh, tap the Refresh button next to your iCloud account in the Reminders section. (You may need to tap Lists at the top left to see your iCloud account.)

Updated reminders are not compatible with Reminders in earlier versions of iOS and macOS. See Updating reminders on iOS 13.

Add a reminder

Ask Siri. Say something like "Add artichokes to my menu." Learn how to ask Siri.

Or do the following:

Touch New Reminder then enters the text.

Use the keyboard shortcut toolbar to do any of the following:

Schedule Date and Time: Touch Date & Time, then select when you want to be reminded.

Add a location: Press the Location button, then select where you want to be reminded - for example, when you get home or when you get in the car, you have a Bluetooth connection to your iPhone.

To receive location-based reminders, you must allow Reminders to use the exact location. Go to Settings> Privacy> Location services, turn on Location services, tap Reminders, select While using an app, and then tap Exact location.

To assign a reminder: (Available in shared lists) Tap Person, then select a person from the shared list (including yourself).

Enter a cursor: Tap Mark to mark an important reminder.

Attach a photo or scanned document: Tap Photos, take a new photo, select an existing photo from the photo library, or scan a document.

To add more details to the reminder, tap Edit details, then do one of the following:

Add notes: Enter more information about the reminder in the Check box.

Add a web link: Enter the URL in the URL field. Reminders display the link as a thumbnail that you can tap to go to the site.

You can get a reminder when chatting with someone in Messaging: Turn on "When sending a message" and then select someone from your contact list. The reminder appears the next time you chat with that person in Messaging.

To set priority: Tap Priority, then select an option.

Touch Done.

Tip: With OS X 10.10 or later, you can transfer edited reminders between Mac and iPhone.

Mark the reminder as complete

Tap the blank circle next to the reminder to mark it as finished and hide it.

To hide completed reminders, tap Next and then tap Finish.

Edit multiple reminders at once

Tap Next, tap Select reminders and then select the reminders you want to edit.

Use the buttons at the bottom of the screen to fill in, mark, add, move, assign, or delete selected reminders and dates.

Move or delete reminders
To rearrange reminders in the list: Tap and hold the reminder you want to move, then drag it to a new location.

To create a subtask: Slide the reminder to the right, then tap Indent. Or drag the reminder to another reminder.

If you delete or move a parent task, the subtasks are also deleted or moved. If you are a parent

CHAPTER NINTEEN
TAKE PICTURES WITH YOUR IPHONE CAMERA

Learn how to take great pictures with your iPhone camera. Choose from camera modes such as Photo, Video, Pano, Timed, Slo-mo, and Portrait (supported models). Enhance your photos with camera features like night mode, live photos, filters, and bursts.

Ask Siri. Say something like "Open the camera". Learn how to ask Siri.

Camera in photo mode, with other modes, left and right below the viewfinder. The Flash, Night Mode, Camera Controls, and Live Photo buttons appear at the top of the screen. The photo and video view buttons are located in the lower-left corner. The Capture button is in the bottom center, and the Camera Chooser Back button is in the lower right corner.

Take a picture
Photo is the normal mode you see when you open the camera. You can take still pictures in Photo mode. Swipe left or right to select another model, such as Video, Pan, Timed, Beat, and Portrait.

Touch the home screen or swipe left from the lock screen to open the camera in photo mode.

Press the shutter button or press a volume button to take the picture.

Note: For your safety, a green dot appears at the top right of the screen when using the camera. See Checking Access to Hardware Features.

Turn the flash on or off
On iPhone XS, iPhone XR, and later, press the Flash button to turn the flash on or off. Or tap Camera Controls, then tap Flash below the frame to select Auto, On, or Off.

167

Tap Flash on iPhone X and earlier, then select Auto, On, or Off.

Set a timer

On iPhone XS, iPhone XR, and later, press the camera controls and then the timer button.

Tap Timer on iPhone X and earlier.

Zoom in or out

To zoom in or out on all models, open the Camera app, and tap the screen.

For iPhone models with dual and triple-camera systems, switch between 1x, 2x, 2.5x and .5x to quickly zoom in or out (depending on model). To zoom more precisely, press and hold the zoom controls, and drag the slider to the right or left.

Laugh and selfie

Take a selfie with the front camera in photo or portrait mode (iPhone X and later).

Switch to the front camera by touching the camera selector Rear-face button or the camera selector Rear-facing button (depending on model).

Hold the iPhone in front of you.

Tip: On iPhone 12 and iPhone 11, you can increase the field of view by tapping the arrows within the frame.

Press the shutter button or press a volume button to take the picture.

To create a mirrored selfie that captures the image as you see it inside the camera, go to Settings> Camera and turn on the Mirror Front Camera feature (available on iPhone XS, iPhone XR, and later).

Adjust the camera's focus and exposure

Before taking a picture, the iPhone automatically adjusts focus and exposure, and Face Detection balances exposure between many faces. Follow these steps to manually adjust focus and exposure:

Touch the screen to display the autofocus area and exposure setting.

Touch the location where you want to move the focus area.

Drag the exposure adjustment knob up or down next to the focus area to adjust the exposure.

To lock manual focus and exposure settings for the following images, touch and hold the focus area until AE / AF lock is displayed; tap the screen to unlock the settings.

On iPhone 11 and later, you can fine-tune and lock the exposure for the next photos. Touch Camera Controls, touch Exposure and then move the slider to adjust the exposure. The exposure is locked until the next time the camera is opened. To preserve the exposure control so that it does not reset when you open Camera, select Settings> Camera> Save settings, and then turn on Exposure adjustment.

Take pictures in low light with night mode

For iPhone 12 and iPhone 11, night mode captures details and illuminates your photos in low light. In night mode, the exposure length is determined automatically, but you can experiment with manual control.

For iPhone 12 models, night mode is available on the front camera for selfies, Ultra Wide (0.5x), and Wide (1x) cameras. On iPhone 11 models, night mode is only available for

Select Photo Mode. In low light, night mode automatically turns on: the night mode button at the top of the screen turns yellow, and a number appears next to the night mode button to indicate how many seconds the camera takes to take pictures.

To experiment with night mode, press the night mode button, then use the slider below the frame to select between Auto and Max timers. In the case of Auto, the time is determined automatically; Max spends the longest time. The selected setting is retained by recording the next night mode.

Press the shutter button and hold the camera still to take the picture.

The crosshairs appear in the frame when the iPhone detects movement during shooting - adjust the crosshair to reduce movement and improve shooting.

To stop recording in night mode, press Stop under the slider.

Take a live photo
Live Photo captures what happens just before and after the photo is taken, including sound.

Select Photo Mode.

Press the Live Photo button to turn Live Photos on or off.

Press the shutter button to take the picture.

You can choose to add effects to Live Photos, such as Loop and Bounce. See Edit live photos on the iPhone.

Take a panoramic picture

Use pan mode to capture landscapes or other images that do not fit on the camera screen.

Select Pan mode, then press the shutter button.

Rotate slowly in the direction of the arrow and hold it in the centerline.

Press the shutter button again to finish.

Press the arrow to pan in the opposite direction. To pan vertically, rotate the iPhone in landscape orientation. You can also reverse the direction of the vertical route.

Take a picture with a filter

Select Photo or portrait mode, and then do one of the following:

For iPhone XS, iPhone XR, and later, tap Camera Controls and then Filters.

For iPhone X and earlier, tap Filters at the top of the screen.

During the display, slide the filters to the right or left to preview; press one to select.

You can remove or change the image filter in Images. See Restoring an Edited Image.

Take continuous pictures

Burst mode takes multiple high-speed shots, so you can choose from a variety of shots. You can take continuous pictures with the rear and front camera.

To take quick pictures, slide the shutter button to the left on iPhone XS, iPhone XR, and later. Press and hold the shutter button on iPhone X and earlier.

The counter shows how many shots you have taken.

Lift your finger to stop.

To select the images you want to keep, tap the Burst thumbnail, then tap Select.

Below the thumbnails, gray dots indicate suggested images for preservation.

Tap the circle at the bottom right of each photo you want to save as a photo, then tap Done.

To delete the entire series, tap the thumbnail, then tap Delete.

Tip: To take a series of pictures, press and hold the volume up key. Go to Settings> Camera and turn on Serial Volume Up (available on iPhone XS, iPhone XR, and later).

www.ingramcontent.com/pod-product-compliance
Lightning Source LLC
Chambersburg PA
CBHW060835220526
45466CB00003B/1115